ROUTLEDGE LIBRARY EDITIONS: GEOLOGY

Volume 12

GEOMORPHOLOGICAL FIELD MANUAL

GEOMORPHOLOGICAL FIELD MANUAL

R.V. DACKOMBE AND V. GARDINER

Routledge
Taylor & Francis Group

LONDON AND NEW YORK

First published in 1983 by George Allen & Unwin (Publishers) Ltd

This edition first published in 2020
by Routledge
2 Park Square, Milton Park, Abingdon, Oxon OX14 4RN

and by Routledge
52 Vanderbilt Avenue, New York, NY 10017

Routledge is an imprint of the Taylor & Francis Group, an informa business

© 1983 R.V. Dackombe and V. Gardiner

British Library Cataloguing in Publication Data
A catalogue record for this book is available from the British Library

ISBN: 978-0-367-18559-6 (Set)
ISBN: 978-0-429-19681-2 (Set) (ebk)
ISBN: 978-0-367-27126-8 (Volume 12) (hbk)
ISBN: 978-0-367-27127-5 (Volume 12) (pbk)
ISBN: 978-0-429-29494-5 (Volume 12) (ebk)

Publisher's Note
The publisher has gone to great lengths to ensure the quality of this reprint but points out that some imperfections in the original copies may be apparent.

Disclaimer
The publisher has made every effort to trace copyright holders and would welcome correspondence from those they have been unable to trace.

Geomorphological Field Manual

R. V. Dackombe
V. Gardiner

London
GEORGE ALLEN & UNWIN
Boston   Sydney

George Allen & Unwin (Publishers) Ltd,
40 Museum Street, London WC1A 1LU, UK

George Allen & Unwin (Publishers) Ltd,
Park Lane, Hemel Hempstead, Herts HP2 4TE, UK

Allen & Unwin Inc.,
9 Winchester Terrace, Winchester, Mass 01890, USA

George Allen & Unwin Australia Pty Ltd,
8 Napier Street, North Sydney, NSW 2060, Australia

British Library Cataloguing in Publication Data

Dackombe, R. V.
  Geomorphological field manual.
1. Geomorphology—Technique
I. Title II. Gardiner, V.
551.4'028   GB400.4

ISBN 0-04-551061-X
ISBN 0-04-551062-8   Pbk

Library of Congress Cataloging in Publication Data

Dackombe, R. V. (Roger V.)
  Geomorphological field manual.

Bibliography: p.
Includes index.
1. Geomorphology—Field work—Handbooks, manuals,
etc. I. Gardiner, V. II. Title.
GB400.5.D32   551.4'028   82-6767
ISBN 0-04-551061-X   AACR2
ISBN 0-04-551062-8   (pbk.)

Set in 9 on 11 point Times by Preface Ltd, Salisbury, Wilts.
and printed in Great Britain
by Hazell Watson and Viney Ltd, Aylesbury, Bucks.

Acknowledgements

We are grateful to acknowledge the help, material and constructive criticism contributed by many workers in a wide range of areas of geomorphology during the preparation of this manual. In particular we are indebted to V. R. Baker, R. Beck, G. S. Boulton, D. Brunsden, M. J. Clark, R. U. Cooke, J. C. Doornkamp, E. M. Durrance, A. S. Goudie, K. J. Gregory, C. Hardwick, K. Hewitt, I. P. Jolliffe, B. Juel-Jensen, J. Mattsson, M. J. Selby, G. S. P. Thomas, C. R. Thorne, J. B. Thornes, A. Warren and W. B. Whalley. Finally, thanks to Gill and Paula, without whose help and forbearance this book could not have been written.

The following organizations and individuals are thanked for permission to reproduce figures and tables. Numbers in parentheses refer to text figures unless otherwise stated:

Figure 1.2 reproduced from *Field geology* (F. H. Lahee), copyright © 1931 McGraw-Hill Book Company, used with the permission of McGraw-Hill Book Company; American Association of Petroleum Geologists (1.3); Figures 2.1 and 2.2 reproduced with permission from *Geomorphology and environmental management* (R. U. Cooke & J. C. Doornkamp), published by Oxford University Press 1974; Canadian Society of Petroleum Geologists (2.5); Blackwell Scientific Publications and the Geological Society of London (2.6, 4.1–3, Tables 4.1–3, 6.5–16 & 6.31); National Research Council of Canada and T. J. Blachut (2.7, 8.1); Figure 3.2 and Table 6.21 reproduced by kind permission of the Editor, British Geomorphological Research Group, and Geo Abstracts Ltd; R. Ginsberg and Springer-Verlag Inc., New York (4.6, 7 & 9); Figure 4.11 reproduced from *Introduction to geology*, vol. 1, Figure 259 (Reid & Watson 1968) by kind permission of J. Watson and Macmillan, London and Basingstoke; Figure 4.12 reproduced from *Measuring stratigraphic sections* by permission of F. E. Kottlowski and Holt, Rinehart & Winston Inc.; T. Meidav and the Society of Exploration Geophysicists (5.2–4); E. Woollard, A. C. Vine and the Geological Society of America (5.5); Figure 5.10 reproduced from *Electrical methods in geophysical prospecting* (Keller & Frischknecht 1966) by kind permission of G. V. Keller, F. Frischknecht and Pergamon Press Ltd; Figure 5.13 reproduced from *Applied geophysics for engineers and geologists* (Griffiths & King 1965) by kind permission of Pergamon Press Ltd; Cambridge University Press (5.14 & 15, Tables 5.6 & 7); Table 5.3 reproduced from *Geophysical Prospecting* 3, 390, by kind permission of E. W. Carpenter and Blackwell Scientific Publications Ltd; J. W. Bray and the Institute of Mining and Metallurgy (6.1, 11.2, Tables 11.1, 11.3–6); Figure 6.2 reproduced with kind permission from *Sedimentary rocks*, 3rd edn, Figures 3–5 (p. 33), by F. J. Pettijohn, copyright 1949, © 1957 by Harper and Row Publishers Inc., copyright © 1975 by F. J. Pettijohn; R. L. Folk (6.3 & 7); Figure 6.3 reproduced from R. L. Folk, *J. Geol.*, copyright © 1954 by The University of Chicago; Figures 6.4 and 6.6 reproduced from *Sand and sandstone* (F. J. Pettijohn, P. E. Potter & R. Seiver 1972) by kind permission of the authors and Springer-Verlag, Heidelberg; Society of Economic Paleontologists and Mineralogists (6.5, Tables 6.32, 7.13a & b); W. C. Krumbein

viii ACKNOWLEDGEMENTS

(6.5); Figure 6.7 reproduced from Sneed & Folk, *J. Geol.*, copyright © 1958 by the University of Chicago; Figure 6.8 reproduced from *Fabric and analysis of soils* (Brewer 1964) by kind permission of R. Brewer and John Wiley and Sons Ltd; Figures 6.9 and 6.11 reproduced with permission from *Stratigraphy and sedimentation*, 2nd edn, by W. C. Krumbein and L. L. Sloss, copyright © 1963 W. H. Freeman & Co.; Elsevier Scientific Publishing Company (6.10, 12, 13, 15 & 16, Tables 6.33 &34); Exxon Production Research Co. (6.10); Figures 6.12 and 6.13, Tables 6.33 and 6.34 reproduced from *Sedimentology* (1963) by kind permission of J. R. L. Allen; Figure 6.14 adapted with kind permission from *Current ripples* (J. R. L. Allen 1968), copyright North Holland Publishing Co.; Figure 6.15 reproduced from *Sedimentology* (1967) by kind permission of W. F. Tanner; The Director, Soil Survey of England and Wales (Tables 6.2–4, 6.22 & 6.24–9, Fig. 13.2); M. J. Selby and Gebrüder Borntraeger (Tables 6.17–20); US Department of Agriculture (Table 6.23); A. V. Jopling and R. G. Walker (Table 6.32); M. de Vries (7.1, 12.2); British Standards Institution (7.2); McGraw-Hill (7.3, 7.7–9, Tables 7.3, & 4); Figure 7.5 reproduced from Flow resistance in gravel-bed rivers (R. D. Hey 1979) *Proc. Am. Soc. Civ. Engrs, J. Hydraul. Div.* **105**, **HY-4**, 365–79, by permission of the American Society of Civil Engineers; material in Table 7.5 from BS 3681; Part 2: 1973 is reproduced by permission of the British Standards Institution, 2 Park Street, London W1A 2BS, from whom complete copies of the standard can be obtained; Figure 7.6 reproduced with permission from US Geological Survey, Water Supply paper 1498-E (Simons & Richardson 1962); F. M. Henderson (7.10); Figures 7.11 and 11.1, and Tables 11.3–6 reproduced from *Hillslope form and process* (M. A. Carson & M. T. Kirkby 1972) by kind permission of Cambridge University Press; H. W. Shen (7.12); Figure 7.14 reproduced from *Geol Soc. Am. Bull.* 1975, by kind permission of V. R. Baker and the Geological Society of America; B. Finlayson (7.15); Figure 8.2 and Table 8.1 reproduced from *The physics of glaciers* (Paterson 1969), by kind permission of Pergamon Press Ltd; J. F. Nye and the International Glacialogical Society (8.3); H. U. Gubler and the International Association of Hydrological Science (IAHS) (8.4); Tables 8.4–7 reproduced by permission of Unesco, IAHS and the World Meteorological Organization from Technical Papers in Hydrology no. 2, *Seasonal snow cover*, © 1970 Unesco/IAHS/WMO; R. Greeley (9.1); R. U. Cooke (9.2, 6 and 7); R. A. Bagnold (9.3); Figures 9.5, 9.6 and Table 9.2 reproduced with permission from US Geological Survey, Professional Paper 1052, *Global sand seas* (E. D. McKee 1979); C. A. M. King (10.2); Hutchinson Publishing Group (10.3); M. Darbyshire (10.5–9); R. Ippen (10.10 & 11); Figures 10.12–15 reproduced from Sternberg, R. W., Predicting initial motion and bedload transport, in D. J. P. Swift *et al.* (eds) *Shelf sediment transport*, Figs 24, 25 and 26, copyright © 1972 by Dowden, Hutchinson & Ross Inc., Stroudsburg, Pa and reprinted by permission of the publishers; D. Brunsden (11.3); Figure 11.4 reprinted with permission from Scott, *Principles of soil mechanics* Figure 9.20, copyright © 1963 Addison-Wesley, Reading, MA – reprinted with permission; A. W. Skempton (11.5); A. Roberts (11.6); Tables 11.4–7 reproduced from *Physical and geotechnical properties of soils* by J. E. Bowles, copyright © 1979 McGraw-Hill Book Company, used with the permission of McGraw-Hill Book Company; Indian Society of Earth Sciences (12.1); British Standards Institution (Table 12.2); The Director, Soil Survey of England and Wales (13.2); Oliver & Boyd (Tables 13.2).

R. V. D.
V. G.

Contents

List of tables

Introduction

This book is not a text on geomorphological techniques. It is intended as a field handbook to be used during the conduct of geomorphological research. Its prime purpose is to act as an *aide memoire* containing information essential for efficient fieldwork. One fundamental assumption made in compiling this manual is that the user is familiar with at least the essentials of the techniques that he proposes to employ. However, where possible sufficient information has been included to refresh the user's memory on the basics of the various techniques. For fuller discussions on the relative merits of a wide range of geomorphological techniques and for further references on them, the user should consult appropriate texts such as Goudie (1981) or the series of technical bulletins produced by the British Geomorphological Research Group.

One of the stimuli that led us to the preparation of this manual was our realisation that, prior to any period of fieldwork, considerable effort was being repeatedly expended in collecting together the materials required in the form of mapping symbols, keys, tables, graphs and so on. Ever present were the risks that a crucial item would be omitted and that, in preparing for closely defined and specific tasks, opportunities would arise that were unforeseen at the time of planning and which therefore could not be capitalised upon. A second stimulus was a feeling that a greater degree of standardisation in recording field information would be beneficial.

The book was initially envisaged as a compilation of material designed to cater for our own personal requirements, but has been extended in order to cover as full a range of geomorphological activities as possible. However, in order to keep it an 'anorak pocket' book, much has, of necessity, been omitted.

Substantial parts of this manual are concerned with quantitative relationships, and therefore formulae are included where appropriate. It is assumed that the user normally has access to a scientific calculator. However, not all calculators are readily capable of computing things such as fractional exponents; the means to accomplish such tasks are included where appropriate. Some nomograms are also included for three further reasons. First, the user may be reluctant to take the calculator into the field; secondly, calculator malfunctions are not entirely unknown; and thirdly, some computations are rather cumbersome and require repeated solution. This is more readily achieved with a simple nomogram than with a non-programmable calculator. It must be noted that such nomograms give only approximate solutions and should be used only for preliminary interpretation and not for final analysis.

Some preliminary analysis of data after or during a day's fieldwork is often necessary in order to plan further fieldwork. Tables of constants, typical values of parameters and some relationships are included to facilitate this. Since much equipment still employs non-SI units, conversions between common units are also given. All formulae have been recast to employ SI units where this was necessary and acceptable. The value of the standard acceleration due to gravity (g) has been taken throughout to be 9.80665 m s^{-2}, and is not given for each individual occurrence.

Specifically excluded from the manual are information on laboratory testing and analysis and advice on the choice and purchase of equipment. Both of these topics are considered in Goudie (1981). Mathematical and statistical tables have also been excluded but the user should have access to appropriate tables, particularly if he does not have access to a calculator.

The structure of the manual follows that of Goudie's (1981) text as far as possible in order to facilitate cross referencing. This does, however, mean that some related items occur widely separated in this text – but they are appropriately cross referenced. Chapters 1–3 deal with the depiction and measurement of landform. The next three chapters deal with the exploration, mapping, recording and classification of landform materials. Chapters 7–11 cover fieldwork concerned with the principal geomorphological processes. Sampling is covered in Chapter 12. Finally, miscellaneous information on such diverse topics as photography, first aid and units of measurement is given in Chapter 13.

1 Topographic survey

This chapter contains materials for determining altitude and differences in elevation only. The user who is concerned primarily with the determination of slope angles should refer to Chapter 3. Most mapping tasks undertaken by the geomorphologist can be resolved into two aspects: first, the provision of a number of fixed points in plan; secondly, the addition of altitudinal information. The former task is essentially a problem in plane geometry and the simple trigonometric relations that may be required to solve such problems can be found in Chapter 13; the latter task is covered in this chapter.

Section 1.1 covers the conversion of apparent slant distances and angles of elevation or declination to a true horizontal distance and an altitude difference. The angles of elevation may be determined by theodolite or inclinometer, and the slant distance may be measured tacheometrically or directly by tape or chain. Section 1.2 contains details of some means of correction for temperature variation in barometric surveying. Finally, Table 1.3 gives a simple guide to map scales and mapping resolution.

1.1 Conversion of apparent slant distances to true horizontal and vertical distances

Tables 1.1 and 1.2 may be used to convert apparent slant distances to true horizontal distance and vertical height difference (Fig. 1.1). Table 1.1 sets out values of $100 \cos^2 \theta$ for use in the formula

$$AC = 100pq \cos^2 \theta$$

Table 1.1 True horizontal distance from stadia intercepts.

Angle °	'	Apparent slant range								
		100	200	300	400	500	600	700	800	900
0	0	100.00	200.00	300.00	400.00	500.00	600.00	700.00	800.00	900.00
0	20	100.00	199.99	299.99	399.99	499.98	599.98	699.98	799.97	899.97
0	40	99.99	199.97	299.96	399.95	499.93	599.92	699.91	799.89	899.88
1	0	99.97	199.94	299.91	399.88	499.85	599.82	699.79	799.76	899.73
1	20	99.95	199.89	299.84	399.78	499.73	599.68	699.62	799.57	899.51
1	40	99.92	199.83	299.75	399.66	499.58	599.49	699.41	799.32	899.24
2	0	99.88	199.76	299.63	399.51	499.39	599.27	699.15	799.03	898.90
2	20	99.83	199.67	299.50	399.34	499.17	599.01	698.84	798.67	898.51
2	40	99.78	199.57	299.35	399.13	498.92	598.70	698.48	798.27	898.05
3	0	99.73	199.45	299.18	398.90	498.63	598.36	698.08	797.81	897.53

Table 1.1 – *continued.*

Angle		100	200	300	400	500	600	700	800	900
°	′					*Apparent slant range*				
3	20	99.66	199.32	298.99	398.65	498.31	597.97	697.63	797.30	896.96
3	40	99.59	199.18	298.77	398.36	497.96	597.55	697.14	796.73	896.32
4	0	99.51	199.03	298.54	398.05	497.57	597.08	696.59	796.11	895.62
4	20	99.43	198.86	298.29	397.72	497.15	596.57	696.00	795.43	894.86
4	40	99.34	198.68	298.01	397.35	496.69	596.03	695.37	794.70	894.04
5	0	99.24	198.48	297.72	396.96	496.20	595.44	694.68	793.92	893.16
5	20	99.14	198.27	297.41	396.54	495.68	594.82	693.95	793.09	892.22
5	40	99.03	198.05	297.08	396.10	495.13	594.15	693.18	792.20	891.23
6	0	98.91	197.81	296.72	395.63	494.54	593.44	692.35	791.26	890.17
6	20	98.78	197.57	296.35	395.13	493.92	592.70	691.48	790.26	889.05
6	40	98.65	197.30	295.96	394.61	493.26	591.91	690.57	789.22	887.87
7	0	98.51	197.03	295.54	394.06	492.57	591.09	689.60	788.12	886.63
7	20	98.37	196.74	295.11	393.48	491.85	590.22	688.60	786.97	885.34
7	40	98.22	196.44	294.66	392.88	491.10	589.32	687.54	785.76	883.98
8	0	98.06	196.13	294.19	392.25	490.32	588.38	686.44	784.50	882.57
8	20	97.90	195.80	293.70	391.60	489.50	587.40	685.30	783.20	881.10
8	40	97.73	195.46	293.19	390.92	488.65	586.38	684.11	781.84	879.56
9	0	97.55	195.11	292.66	390.21	487.76	585.32	682.87	780.42	877.98
9	20	97.37	194.74	292.11	389.48	486.85	584.22	681.59	778.96	876.33
9	40	97.18	194.36	291.54	388.72	485.90	583.08	680.26	777.44	874.62
10	0	96.98	193.97	290.95	387.94	484.92	581.91	678.89	775.88	872.86
10	20	96.78	193.56	290.35	387.13	483.91	580.69	677.48	774.26	871.04
10	40	96.57	193.15	289.72	386.30	482.87	579.44	676.02	772.59	869.17
11	0	96.36	192.72	289.08	385.44	481.80	578.16	674.51	770.87	867.23
11	20	96.14	192.28	288.41	384.55	480.69	576.83	672.97	769.10	865.24
11	40	95.91	191.82	287.73	383.64	479.55	575.46	671.38	767.29	863.20
12	0	95.68	191.35	287.03	382.71	478.39	574.06	669.74	765.42	861.10
12	20	95.44	190.88	286.31	381.75	477.19	572.63	668.06	763.50	858.94
12	40	95.19	190.38	285.58	380.77	475.96	571.15	666.34	761.53	856.73
13	0	94.94	189.88	284.82	379.76	474.70	569.64	664.58	759.52	854.46
13	20	94.68	189.36	284.04	378.73	473.41	568.09	662.77	757.45	852.13
13	40	94.42	188.84	283.25	377.67	472.09	566.51	660.92	755.34	849.76
14	0	94.15	188.29	282.44	376.59	470.74	564.88	659.03	753.18	847.33
14	20	93.87	187.74	281.61	375.49	469.36	563.23	657.10	750.97	844.84
14	40	93.59	187.18	280.77	374.36	467.95	561.54	655.12	748.71	842.30
15	0	93.30	186.60	279.90	373.21	466.51	559.81	653.11	746.41	839.71
15	20	93.01	186.01	279.02	372.03	465.04	558.04	651.05	744.06	837.07
15	40	92.71	185.42	278.12	370.83	463.54	556.25	648.95	741.66	834.37
16	0	92.40	184.80	277.21	369.61	462.01	554.41	646.82	739.22	831.62
16	20	92.09	184.18	276.27	368.36	460.46	552.55	644.64	736.73	828.82
16	40	91.77	183.55	275.32	367.10	458.87	550.65	642.42	734.20	825.97
17	0	91.45	182.90	274.36	365.81	457.26	548.71	640.16	731.62	823.07
17	20	91.12	182.25	273.37	364.50	455.62	546.74	637.87	728.99	820.11
17	40	90.79	181.58	272.37	363.16	453.95	544.74	635.53	726.32	817.11
18	0	90.45	180.90	271.35	361.80	452.25	542.71	633.16	723.61	814.06
18	20	90.11	180.21	270.32	360.42	450.53	540.64	630.74	720.85	810.96
18	40	89.76	179.51	269.27	359.02	448.78	538.54	628.29	718.05	807.80
19	0	89.40	178.80	268.20	357.60	447.00	536.40	625.80	715.20	804.60
19	20	89.04	178.08	267.12	356.16	445.20	534.24	623.28	712.32	801.36
19	40	88.67	177.35	266.02	354.69	443.37	532.04	620.72	709.39	798.06
20	0	88.30	176.60	264.91	353.21	441.51	529.81	618.12	706.42	794.72
20	20	87.93	175.85	263.78	351.70	439.63	527.55	615.48	703.41	791.33
20	40	87.54	175.09	262.63	350.18	437.72	525.26	612.81	700.35	787.90
21	0	87.16	174.31	261.47	348.63	435.79	522.94	610.10	697.26	784.42
21	20	86.77	173.53	260.30	347.06	433.83	520.59	607.36	694.12	780.89
21	40	86.37	172.74	259.11	345.47	431.84	518.21	604.58	690.95	777.32
22	0	85.97	171.93	257.90	343.87	429.83	515.80	601.77	687.74	773.70
22	20	85.56	171.12	256.68	342.24	427.80	513.36	598.92	684.48	770.04
22	40	85.15	170.30	255.45	340.60	425.75	510.89	596.04	681.19	766.34
23	0	84.73	169.47	254.20	338.93	423.66	508.40	593.13	677.86	762.60

23	20	84.31	168.62	252.94	337.25	421.56	505.87	590.18	674.50	758.81
23	40	83.89	167.77	251.66	335.55	419.43	503.32	587.21	671.09	754.98
24	0	83.46	166.91	250.37	333.83	417.28	500.74	584.20	667.65	751.11
24	20	83.02	166.04	249.07	332.09	415.11	498.13	581.15	664.18	747.20
24	40	82.58	165.17	247.75	330.33	412.91	495.50	578.08	660.66	743.25
25	0	82.14	164.28	246.42	328.56	410.70	492.84	574.98	657.12	739.25
25	20	81.69	163.38	245.07	326.77	408.46	490.15	571.84	653.53	735.22
25	40	81.24	162.48	243.72	324.96	406.20	487.44	568.68	649.92	731.15
26	0	80.78	161.57	242.35	323.13	403.92	484.70	565.48	646.26	727.05
26	20	80.32	160.65	240.97	321.29	401.61	481.94	562.26	642.58	722.90
26	40	79.86	159.72	239.57	319.43	399.29	479.15	559.01	638.86	718.72
27	0	79.39	158.78	238.17	317.56	396.95	476.34	555.72	635.11	714.50
27	20	78.92	157.83	236.75	315.67	394.58	473.50	552.42	631.33	710.25
27	40	78.44	156.88	235.32	313.76	392.20	470.64	549.08	627.52	705.96
28	0	77.96	155.92	233.88	311.84	389.80	467.76	545.72	623.68	701.64
28	20	77.48	154.95	232.43	309.90	387.38	464.85	542.33	619.80	697.28
28	40	76.99	153.98	230.96	307.95	384.94	461.93	538.91	615.90	692.89
29	0	76.50	152.99	229.49	305.98	382.48	458.98	535.47	611.97	688.46
29	20	76.00	152.00	228.00	304.00	380.00	456.00	532.01	608.01	684.01
29	40	75.50	151.00	226.51	302.01	377.51	453.01	528.51	604.02	679.52
30	0	75.00	150.00	225.00	300.00	375.00	450.00	525.00	600.00	675.00
30	20	74.49	148.99	223.48	297.98	372.47	446.97	521.46	595.96	670.45
30	40	73.99	147.97	221.96	295.94	369.93	443.91	517.90	591.89	665.87
31	0	73.47	146.95	220.42	293.89	367.37	440.84	514.32	587.79	661.26
31	20	72.96	145.92	218.87	291.83	364.79	437.75	510.71	583.67	656.62
31	40	72.44	144.88	217.32	289.76	362.20	434.64	507.08	579.52	651.96
32	0	71.92	143.84	215.76	287.67	359.59	431.51	503.43	575.35	647.27
32	20	71.39	142.79	214.18	285.58	356.97	428.37	499.76	571.15	642.55
32	40	70.87	141.73	212.60	283.47	354.33	425.20	496.07	566.94	637.80
33	0	70.34	140.67	211.01	281.35	351.68	422.02	492.36	562.69	633.03
33	20	69.80	139.61	209.41	279.22	349.02	418.82	488.63	558.43	628.24
33	40	69.27	138.54	207.81	277.07	346.34	415.61	484.88	554.15	623.42
34	0	68.73	137.46	206.19	274.92	343.65	412.38	481.11	549.84	618.57
34	20	68.19	136.38	204.57	272.76	340.95	409.14	477.33	545.52	613.71
34	40	67.65	135.29	202.94	270.59	338.23	405.88	473.53	541.17	608.82
35	0	67.10	134.20	201.30	268.40	335.51	402.61	469.71	536.81	603.91
35	20	66.55	133.11	199.66	266.21	332.77	399.32	465.87	532.43	598.98
35	40	66.00	132.01	198.01	264.01	330.02	396.02	462.02	528.02	594.03
36	0	65.45	130.90	196.35	261.80	327.25	392.71	458.16	523.61	589.06
36	20	64.90	129.79	194.69	259.59	324.48	389.38	454.28	519.17	584.07
36	40	64.34	128.68	193.02	257.36	321.70	386.04	450.38	514.72	579.06
37	0	63.78	127.56	191.35	255.13	318.91	382.69	446.47	510.25	574.04
37	20	63.22	126.44	189.67	252.89	316.11	379.33	442.55	505.77	569.00
37	40	62.66	125.32	187.98	250.64	313.30	375.96	438.62	501.28	563.94
38	0	62.10	124.19	186.29	248.38	310.48	372.58	434.67	496.77	558.86
38	20	61.53	123.06	184.59	246.12	307.65	369.18	430.72	492.25	553.78
38	40	60.96	121.93	182.89	243.86	304.82	365.78	426.75	487.71	548.68
39	0	60.40	120.79	181.19	241.58	301.98	362.37	422.77	483.16	543.56
39	20	59.83	119.65	179.48	239.30	299.13	358.95	418.78	478.61	538.43
39	40	59.25	118.51	177.76	237.02	296.27	355.53	414.78	474.04	533.29
40	0	58.68	117.36	176.05	234.73	293.41	352.09	410.78	469.46	528.14
40	20	58.11	116.22	174.33	232.44	290.54	348.65	406.76	464.87	522.98
40	40	57.53	115.07	172.60	230.14	287.67	345.21	402.74	460.27	517.81
41	0	56.96	113.92	170.88	227.83	284.79	341.75	398.71	455.67	512.63
41	20	56.38	112.76	169.15	225.53	281.91	338.29	394.67	451.06	507.44
41	40	55.80	111.61	167.41	223.22	279.02	334.83	390.63	446.44	502.24
42	0	55.23	110.45	165.68	220.91	276.13	331.36	386.58	441.81	497.04
42	20	54.65	109.29	163.94	218.59	273.24	327.88	382.53	437.18	491.83
42	40	54.07	108.14	162.20	216.27	270.34	324.41	378.48	432.54	486.61
43	0	53.49	106.98	160.46	213.95	267.44	320.93	374.41	427.90	481.39
43	20	52.91	105.81	158.72	211.63	264.54	317.44	370.35	423.26	476.17
43	40	52.33	104.65	156.98	209.31	261.63	313.96	366.28	418.61	470.94
44	0	51.74	103.49	155.23	206.98	258.72	310.47	362.21	413.96	465.70
44	20	51.16	102.33	153.49	204.65	255.82	306.98	358.14	409.31	460.47
44	40	50.58	101.16	151.75	202.33	252.91	303.49	354.07	404.65	455.24
45	0	50.00	100.00	150.00	200.00	250.00	300.00	350.00	400.00	450.00

Table 1.2 True vertical differences from stadia intercepts.

Angle °	'	Apparent slant range								
		100	200	300	400	500	600	700	800	900
0	0	0.00	0.00	0.00	0.00	0.00	0.00	0.00	0.00	0.00
0	20	.58	1.16	1.75	2.33	2.91	3.49	4.07	4.65	5.24
0	40	1.16	2.33	3.49	4.65	5.82	6.98	8.14	9.31	10.47
1	0	1.74	3.49	5.23	6.98	8.72	10.47	12.21	13.96	15.70
1	20	2.33	4.65	6.98	9.31	11.63	13.96	16.28	18.61	20.94
1	40	2.91	5.81	8.72	11.63	14.54	17.44	20.35	23.26	26.17
2	0	3.49	6.98	10.46	13.95	17.44	20.93	24.41	27.90	31.39
2	20	4.07	8.14	12.20	16.27	20.34	24.41	28.48	32.54	36.61
2	40	4.65	9.29	13.94	18.59	23.24	27.88	32.53	37.18	41.83
3	0	5.23	10.45	15.68	20.91	26.13	31.36	36.58	41.81	47.04
3	20	5.80	11.61	17.41	23.22	29.02	34.83	40.63	46.44	52.24
3	40	6.38	12.76	19.15	25.53	31.91	38.29	44.67	51.06	57.44
4	0	6.96	13.92	20.88	27.83	34.79	41.75	48.71	55.67	62.63
4	20	7.53	15.07	22.60	30.14	37.67	45.21	52.74	60.27	67.81
4	40	8.11	16.22	24.33	32.44	40.54	48.65	56.76	64.87	72.98
5	0	8.68	17.36	26.05	34.73	43.41	52.09	60.78	69.46	78.14
5	20	9.25	18.51	27.76	37.02	46.27	55.53	64.78	74.04	83.29
5	40	9.83	19.65	29.48	39.30	49.13	58.95	68.78	78.61	88.43
6	0	10.40	20.79	31.19	41.58	51.98	62.37	72.77	83.16	93.56
6	20	10.96	21.93	32.89	43.86	54.82	65.78	76.75	87.71	98.68
6	40	11.53	23.06	34.59	46.12	57.65	69.18	80.72	92.25	103.78
7	0	12.10	24.19	36.29	48.38	.60.48	72.58	84.67	96.77	108.86
7	20	12.66	25.32	37.98	50.64	63.30	75.96	88.62	101.28	113.94
7	40	13.22	26.44	39.67	52.89	66.11	79.33	92.55	105.77	119.00
8	0	13.78	27.56	41.35	55.13	68.91	82.69	96.47	110.25	124.04
8	20	14.34	28.68	43.02	57.36	71.70	86.04	100.38	114.72	129.06
8	40	14.90	29.79	44.69	59.59	74.48	89.38	104.28	119.17	134.07
9	0	15.45	30.90	46.35	61.80	77.25	92.71	108.16	123.61	139.06
9	20	16.00	32.01	48.01	64.01	80.02	96.02	112.02	128.02	144.03
9	40	16.55	33.11	49.66	66.21	82.77	99.32	115.87	132.43	148.98
10	0	17.10	34.20	51.30	68.40	85.51	102.61	119.71	136.81	153.91
10	20	17.65	35.29	52.94	70.59	88.23	105.88	123.53	141.17	158.82
10	40	18.19	36.38	54.57	72.76	90.95	109.14	127.33	145.52	163.71
11	0	18.73	37.46	56.19	74.92	93.65	112.38	131.11	149.84	168.57
11	20	19.27	38.54	57.81	77.07	96.34	115.61	134.88	154.15	173.42
11	40	19.80	39.61	59.41	79.22	99.02	118.82	138.63	158.43	178.24
12	0	20.34	40.67	61.01	81.35	101.68	122.02	142.36	162.69	183.03
12	20	20.87	41.73	62.60	83.47	104.33	125.20	146.07	166.94	187.80
12	40	21.39	42.79	64.18	85.58	106.97	128.37	149.76	171.15	192.55
13	0	21.92	43.84	65.76	87.67	109.59	131.51	153.43	175.35	197.27
13	20	22.44	44.88	67.32	89.76	112.20	134.64	157.08	179.52	201.96
13	40	22.96	45.92	68.87	91.83	114.79	137.75	160.71	183.67	206.62
14	0	23.47	46.95	70.42	93.89	117.37	140.84	164.32	187.79	211.26
14	20	23.99	47.97	71.96	95.94	119.93	143.91	167.90	191.89	215.87
14	40	24.49	48.99	73.48	97.98	122.47	146.97	171.46	195.96	220.45
15	0	25.00	50.00	75.00	100.00	125.00	150.00	175.00	200.00	225.00
15	20	25.50	51.00	76.51	102.01	127.51	153.01	178.51	204.02	229.52
15	40	26.00	52.00	78.00	104.00	130.00	156.00	182.01	208.01	234.01
16	0	26.50	52.99	79.49	105.98	132.48	158.98	185.47	211.97	238.46
16	20	26.99	53.98	80.96	107.95	134.94	161.93	188.91	215.90	242.89
16	40	27.48	54.95	82.43	109.90	137.38	164.85	192.33	219.80	247.28
17	0	27.96	55.92	83.88	111.84	139.80	167.76	195.72	223.68	251.64
17	20	28.44	56.88	85.32	113.76	142.20	170.64	199.08	227.52	255.96
17	40	28.92	57.83	86.75	115.67	144.58	173.50	202.42	231.33	260.25
18	0	29.39	58.78	88.17	117.56	146.95	176.34	205.72	235.11	264.50
18	20	29.86	59.72	89.57	119.43	149.29	179.15	209.01	238.86	268.72
18	40	30.32	60.65	90.97	121.29	151.61	181.94	212.26	242.58	272.90
19	0	30.78	61.57	92.35	123.13	153.92	184.70	215.48	246.26	277.05
19	20	31.24	62.48	93.72	124.96	156.20	187.44	218.68	249.92	281.15
19	40	31.69	63.38	95.07	126.77	158.46	190.15	221.84	253.53	285.22

20	0	32.14	64.28	96.42	128.56	160.70	192.84	224.98	257.12	289.25
20	20	32.58	65.17	97.75	130.33	162.91	195.50	228.08	260.66	293.25
20	40	33.02	66.04	99.07	132.09	165.11	198.13	231.15	264.18	297.20
21	0	33.46	66.91	100.37	133.83	167.28	200.74	234.20	267.65	301.11
21	20	33.89	67.77	101.66	135.55	169.43	203.32	237.21	271.09	304.98
21	40	34.31	68.62	102.94	137.25	171.56	205.87	240.18	274.50	308.81
22	0	34.73	69.47	104.20	138.93	173.66	208.40	243.13	277.86	312.60
22	20	35.15	70.30	105.45	140.60	175.75	210.89	246.04	281.19	316.34
22	40	35.56	71.12	106.68	142.24	177.80	213.36	248.92	284.48	320.04
23	0	35.97	71.93	107.90	143.87	179.83	215.80	251.77	287.74	323.70
23	20	36.37	72.74	109.11	145.47	181.84	218.21	254.58	290.95	327.32
23	40	36.77	73.53	110.30	147.06	183.83	220.59	257.36	294.12	330.89
24	0	37.16	74.31	111.47	148.63	185.79	222.94	260.10	297.26	334.42
24	20	37.54	75.09	112.63	150.18	187.72	225.26	262.81	300.35	337.90
24	40	37.93	75.85	113.78	151.70	189.63	227.55	265.48	303.41	341.33
25	0	38.30	76.60	114.91	153.21	191.51	229.81	268.12	306.42	344.72
25	20	38.67	77.35	116.02	154.69	193.37	232.04	270.72	309.39	348.06
25	40	39.04	78.08	117.12	156.16	195.20	234.24	273.28	312.32	351.36
26	0	39.40	78.80	118.20	157.60	197.00	236.40	275.80	315.20	354.60
26	20	39.76	79.51	119.27	159.02	198.78	238.54	278.29	318.05	357.80
26	40	40.11	80.21	120.32	160.42	200.53	240.64	280.74	320.85	360.96
27	0	40.45	80.90	121.35	161.80	202.25	242.71	283.16	323.61	364.06
27	20	40.79	81.58	122.37	163.16	203.95	244.74	285.53	326.32	367.11
27	40	41.12	82.25	123.37	164.50	205.62	246.74	287.87	328.99	370.11
28	0	41.45	82.90	124.36	165.81	207.26	248.71	290.16	331.62	373.07
28	20	41.77	83.55	125.32	167.10	208.87	250.65	292.42	334.20	375.97
28	40	42.09	84.18	126.27	168.36	210.46	252.55	294.64	336.73	378.82
29	0	42.40	84.80	127.21	169.61	212.01	254.41	296.82	339.22	381.62
29	20	42.71	85.42	128.12	170.83	213.54	256.25	298.95	341.66	384.37
29	40	43.01	86.01	129.02	172.03	215.04	258.04	301.05	344.06	387.07
30	0	43.30	86.60	129.90	173.21	216.51	259.81	303.11	346.41	389.71
30	20	43.59	87.18	130.77	174.36	217.95	261.54	305.12	348.71	392.30
30	40	43.87	87.74	131.61	175.49	219.36	263.23	307.10	350.97	394.84
31	0	44.15	88.29	132.44	176.59	220.74	264.88	309.03	353.18	397.33
31	20	44.42	88.84	133.25	177.67	222.09	266.51	310.92	355.34	399.76
31	40	44.68	89.36	134.04	178.73	223.41	268.09	312.77	357.45	402.13
32	0	44.94	89.88	134.82	179.76	224.70	269.64	314.58	359.52	404.46
32	20	45.19	90.38	135.58	180.77	225.96	271.15	316.34	361.53	406.73
32	40	45.44	90.88	136.31	181.75	227.19	272.63	318.06	363.50	408.94
33	0	45.68	91.35	137.03	182.71	228.39	274.06	319.74	365.42	411.10
33	20	45.91	91.82	137.73	183.64	229.55	275.46	321.38	367.29	413.20
33	40	46.14	92.28	138.41	184.55	230.69	276.83	322.97	369.10	415.24
34	0	46.36	92.72	139.08	185.44	231.80	278.16	324.51	370.87	417.23
34	20	46.57	93.15	139.72	186.30	232.87	279.44	326.02	372.59	419.17
34	40	46.78	93.56	140.35	187.13	233.91	280.69	327.48	374.26	421.04
35	0	46.98	93.97	140.95	187.94	234.92	281.91	328.89	375.88	422.86
35	20	47.18	94.36	141.54	188.72	235.90	283.08	330.26	377.44	424.62
35	40	47.37	94.74	142.11	189.48	236.85	284.22	331.59	378.96	426.33
36	0	47.55	95.11	142.66	190.21	237.76	285.32	332.87	380.42	427.98
36	20	47.73	95.46	143.19	190.92	238.65	286.38	334.11	381.84	429.56
36	40	47.90	95.80	143.70	191.60	239.50	287.40	335.30	383.20	431.10
37	0	48.06	96.13	144.19	192.25	240.32	288.38	336.44	384.50	432.57
37	20	48.22	96.44	144.66	192.88	241.10	289.32	337.54	385.76	433.98
37	40	48.37	96.74	145.11	193.48	241.85	290.22	338.60	386.97	435.34
38	0	48.51	97.03	145.54	194.06	242.57	291.09	339.60	388.12	436.63
38	20	48.65	97.30	145.96	194.61	243.26	291.91	340.57	389.22	437.87
38	40	48.78	97.57	146.35	195.13	243.92	292.70	341.48	390.26	439.05
39	0	48.91	97.81	146.72	195.63	244.54	293.44	342.35	391.26	440.17
39	20	49.03	98.05	147.08	196.10	245.13	294.15	343.18	392.20	441.23
39	40	49.14	98.27	147.41	196.54	245.68	294.82	343.95	393.09	442.22
40	0	49.24	98.48	147.72	196.96	246.20	295.44	344.68	393.92	443.16
40	20	49.34	98.68	148.01	197.35	246.69	296.03	345.37	394.70	444.04
40	40	49.43	98.86	148.29	197.72	247.15	296.57	346.00	395.43	444.86
41	0	49.51	99.03	148.54	198.05	247.57	297.08	346.59	396.11	445.62
41	20	49.59	99.18	148.77	198.36	247.96	297.55	347.14	396.73	446.32
41	40	49.66	99.32	148.99	198.65	248.31	297.97	347.63	397.30	446.96
42	0	49.73	99.45	149.18	198.90	248.63	298.36	348.08	397.81	447.53
42	20	49.78	99.57	149.35	199.13	248.92	298.70	348.48	398.27	448.05

Table 1.2 – *continued.*

Angle		*Apparent slant range*							
° ′	*100*	*200*	*300*	*400*	*500*	*600*	*700*	*800*	*900*
42 40	49.83	99.67	149.50	199.34	249.17	299.01	348.84	398.67	448.51
43 0	49.88	99.76	149.63	199.51	249.39	299.27	349.15	399.03	448.90
43 20	49.92	99.83	149.75	199.66	249.58	299.49	349.41	399.32	449.24
43 40	49.95	99.89	149.84	199.78	249.73	299.68	349.62	399.57	449.51
44 0	49.97	99.94	149.91	199.88	249.85	299.82	349.79	399.76	449.73
44 20	49.99	99.97	149.96	199.95	249.93	299.92	349.91	399.89	449.88
44 40	50.00	99.99	149.99	199.99	249.98	299.98	349.98	399.97	449.97
45 0	50.00	100.00	150.00	200.00	250.00	300.00	350.00	400.00	450.00

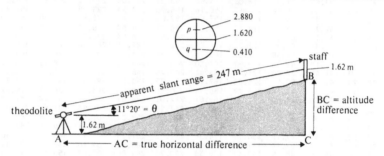

Figure 1.1   The use of the theodolite to determine apparent slant range, horizontal distance and altitude difference.

and Table 1.2 sets out values of $100 \sin^2(\theta/2)$ for use in the formula

$$BC = 100pq \sin^2(\theta/2),$$

where AC = true horizontal distance, BC = altitude difference, pq = difference between stadia readings, θ = angle of elevation or declination. The values given in the tables are for 20′ intervals of arc and where required linear interpolation may be used between the tabulated values. An example of the use of the tables is given below and is illustrated in Figure 1.1.

The height of the theodolite axis at station A is 1.62 m and a sight is made onto the staff at B with the crosswire at the 1.62 m mark.

upper stadia	= 2.880
lower stadia	= 0.410
difference in stadia readings *(pq)* =	2.47

Therefore apparent slant range = 247 m.

A is lower than B and the angle of elevation is 11˘20′. Hence, from Table 1.1 at 11°20′

$$200 \text{ m} = 192.28$$
$$40 \text{ m} = 38.46$$
$$7 \text{ m} = 6.73$$
$$\overline{237.47 \text{ m}}$$

Therefore, true horizontal distance (AC) = 237.47 m.
  From Table 1.2 at 11°20′

$$200 \text{ m} = 38.54$$
$$40 \text{ m} = 7.71$$
$$7 \text{ m} = 1.35$$
$$\overline{47.60 \text{ m}}$$

Therefore, difference in altitude (BC) = 47.60 m.

1.2 Correction for temperature in barometric (aneroid or altimeter) surveying

Barometric pressure varies with both altitude and temperature and a correction has to be applied to barometric readings if temperatures differ between the observing stations. Figure 1.2 allows a correction to be made for the difference in temperature between each pair of stations and the algebraic sum of these corrections is carried forward from station to station. This method is most appropriate for areas of high relief where differences in temperature are considerable. The measured differences in elevation between the two stations and the sum of temperatures at the two stations are located on the left- and right-hand bars of the nomogram respectively and are joined by a straight-edge. The correction to be applied to the elevation difference may then be read off·from where the straight-edge crosses the central bar. This correction is added to the observed value of elevation if the sum of temperatures is over 37.8 °C, and subtracted from it if the sum is less than 37.8 °C. The nomogram may also be used to compare individual values with that of a selected datum altitude.

  Figure 1.3 allows a similar correction to be applied. However, it is less accurate, although quicker in use, and is most suited to areas of moderate relief. The value of absolute difference in altitude between the datum elevation and the observation point is located on the vertical scale. The correction for temperature for the reading is then found by starting at the appropriate altitude difference on the vertical axis and reading off the correction value where the difference value and the appropriate temperature of the observation point intersect. This point of intersection will

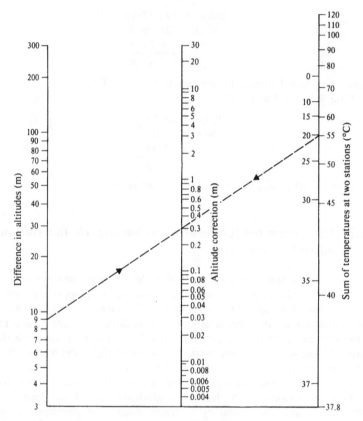

Figure 1.2 Nomogram for application of temperature correction to barometric readings (after Lahee 1931).

generally fall between two of the curved lines, and the figure that falls between the lines is the correction value in metres. If the datum value is greater than the observed value the correction is subtracted, if the datum is less than the observed value the correction is added.

Table 1.3 Guide to map scales and mapping resolution.

Map scale	Feet on ground per inch on map	Metres on ground per centimetre on map	Inches on map per mile on ground	Centimetres on map per kilometre on ground	True ground width of map lines of various thicknesses (mm). Ground widths are given in metres					
					0.10	0.13	0.25	0.35	0.50	1.00
1 : 500	41.67	5.00	126.72	200.00	0.05	0.07	0.13	0.18	0.25	0.50
1 : 600	50.00	6.00	105.60	166.67	0.06	0.08	0.15	0.20	0.30	0.60
1 : 1000	83.33	10.00	63.36	100.00	0.10	0.13	0.25	0.35	0.50	1.00
1 : 1250	104.17	12.50	50.68	80.00	0.13	0.16	0.31	0.44	0.63	1.25
1 : 1500	125.00	15.00	42.24	66.67	0.15	0.20	0.38	0.53	0.75	1.50
1 : 2000	166.67	20.00	31.68	50.00	0.20	0.26	0.50	0.70	1.00	2.00
1 : 2500	208.33	25.00	25.34	40.00	0.25	0.33	0.63	0.88	1.25	2.50
1 : 3000	250.00	30.00	21.12	33.33	0.30	0.39	0.75	1.05	1.50	3.00
1 : 5000	416.67	50.00	12.67	20.00	0.50	0.65	1.25	1.75	2.50	5.00
1 : 6000	500.00	60.00	10.56	16.67	0.60	0.78	1.50	2.10	3.00	6.00
1 : 10 000	833.33	100.00	6.34	10.00	1.00	1.30	2.50	3.50	5.00	10.00
1 : 10 560	880.00	105.60	6.00	9.47	1.06	1.37	2.64	3.70	5.28	10.56
1 : 12 000	1000.00	120.00	5.28	8.33	1.20	1.56	3.00	4.20	6.00	12.00
1 : 20 000	1666.67	200.00	3.17	5.00	2.00	2.60	5.00	7.00	10.00	20.00
1 : 24 000	2000.00	240.00	2.64	4.17	2.40	3.12	6.00	8.40	12.00	24.00
1 : 25 000	2083.33	250.00	2.53	4.00	2.50	3.25	6.25	8.75	12.50	25.00
1 : 40 000	3333.33	400.00	1.58	2.50	4.00	5.20	10.00	14.00	20.00	40.00
1 : 48 000	4000.00	480.00	1.32	2.08	4.80	6.24	12.00	16.80	24.00	48.00
1 : 50 000	4166.67	500.00	1.27	2.00	5.00	6.50	12.50	17.50	25.00	50.00
1 : 63 360	5280.00	633.60	1.00	1.58	6.34	8.24	15.84	22.18	31.68	63.36

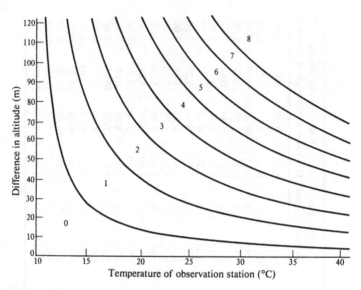

Figure 1.3 Chart for estimation of temperature correction for barometric surveying (based upon Rogers 1947).

2 Geomorphological mapping

This chapter includes material for use in mapping landforms. The first section is a simple set of symbols for producing a morphological map. The second is a moderately comprehensive set of symbols for general-purpose geomorphological mapping, in which genesis of landforms is considered. The final section gives sets of symbols for more specialised applications. (See Goudie 1981, pp. 66–78.)

2.1 Basic morphological mapping symbols

The set of symbols shown in Figure 2.1 is designed for the depiction of landform by recognition of units of equal slope (facets), delimited by either breaks or changes of slope. It is based on schemes suggested by Waters (1958) and Savigear (1965), as presented by Cooke and Doornkamp (1974).

2.2 Geomorphological (genetic) mapping symbols

The symbols in Figure 2.2, which are based largely upon those in Cooke and Doornkamp (1974), are intended for general-purpose geomorphological mapping in which landform genesis is considered. Different processes of origin may be indicated by colours, and suggested colours are shown in brackets for each group of symbols. If more detailed and elaborate schemes of symbols are required for general-purpose mapping, reference should be made to Demek (1972) for large-scale maps and to Demek and Embleton (1978) for medium-scale maps. Symbols for use in more specialised applications are shown in the final section of this chapter.

2.3 Symbols for more detailed mapping

This section includes symbols for use in more detailed mapping in particular process environments. It gives more extensive sets of symbols than

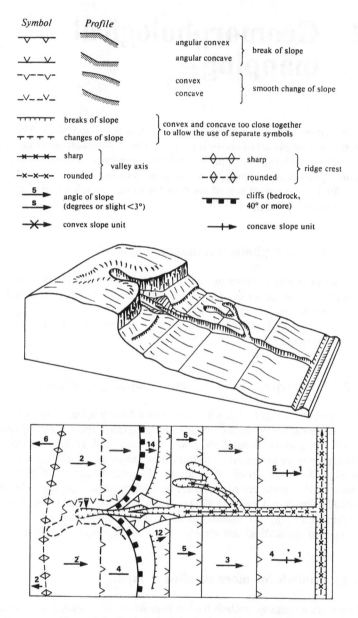

Symbol	Profile	
⌄ ⌄		angular convex
⌄ ⌄		angular concave } break of slope
⁻⌄⁻⁻⌄⁻		convex
₋⌄₋ ₋⌄₋		concave } smooth change of slope

breaks of slope
changes of slope } convex and concave too close together to allow the use of separate symbols

—×—×—×— sharp
—×—×—×— rounded } valley axis

◇—◇ sharp
◇—◇ rounded } ridge crest

→ 5 angle of slope (degrees or slight <3°)

cliffs (bedrock, 40° or more)

convex slope unit   concave slope unit

Figure 2.1 Basic morphological mapping symbols.

Glacial and periglacial features (light blue)

	snowfield		lateral moraine
	glacier ice		medial moraine
	cirque		esker
	rock bar		kame deposits
	glacial trough		dead ice depression
	hanging glacial valley		glacial outwash
	avalanche track		thaw basin
	large glacial drainage channel		thaw subsidence
	drumlin		pingo
	roches moutonnées		patterned ground
	ground moraine		stone stripes
	terminal moraine		small glacial drainage channel

Forms of fluvial origin (dark blue)

	stream		alluvial fan
	river channel		delta
	dry river channel		swamp
	waterfall		crevasse
	rapids		permanent lake
	plunge pool		temporary lake
	sand bar		area susceptible to flooding
	cut-off meander		dry valley
	oxbow lake		asymmetric valley
	sedimentation boundary		bank, cut
	boundary of valley floor		bank, tree-lined
	levees		bank, protected
	point bars		water seepage line
	erosion terrace		low relief depression
	depositional terrace		

Karst features (orange)

	conical karst		doline
	tower karst		swallow hole
	labyrinth karst		cave
	limestone pavement		gorge
	clints and grikes		

Slope instability features (brown)

	landslide (type undetermined)		sand run

Figure 2.2 Geomorphological (genetic) mapping symbols·(based upon Cooke & Doornkamp 1974).

rotational slide (with back tilt)
non-circular rotational slide with graben
flow slide
mud slide

rock fall
solifluction lobe
soil creep

Man-made features (black)

quarry
sand pit
gravel pit
mining subsidence
tumuli, etc.
mining pits
tips (mounds)

filled hollows
transport route
leat
embankment
surface workings
settlement area
surface heavily remodelled

Features resulting from bedrock structure (purple)

escarpment (cuesta scarp)
dipslope
fault scarp

fault-line scarp
broad anticlinal crest
broad synclinal depression

Features of volcanic origin (red)

crater
ash cone
cinder cone
caldera
lava field or flow
block lava

ropy lava
pillow lava
volcanic plug
cinder field
dyke
geyser

Coastal features (green)

surf zone
prevailing drift
rock shore platform
mangrove or algal peat
prominent } beach ridges complex
traces
cliff face
abandoned cliff face
valleys left hanging by cliff recession (valley cross section shown)

sea cave
accretion zone
salt pan
offshore bar
spit
beach ridges
lagoon
stack
undifferentiated coastal plain and sabkha surface

Aeolian features (yellow)

wind-blown sand plain
plain with desert pavement
foredune

$\hat{H}$ high
$\hat{M}$ medium } dunes
$\hat{L}$ low

Figure 2.2 – *continued*.

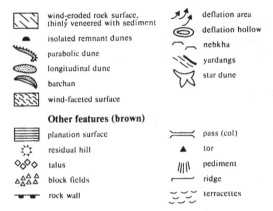

Figure 2.2 – *continued.*

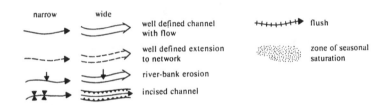

Figure 2.3 Symbols for mapping channel typology (based upon Gregory 1979).

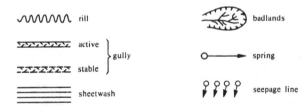

Figure 2.4 Symbols for mapping fluvial erosion on slopes.

those in Figure 2.2, but only for limited contexts and purposes. Symbols for mapping some aspects of fluvial landforms are shown in Figures 2.3, 2.4 and 2.5. Figure 2.3 is for depiction of channel typology, after Gregory (1979). Fluvial erosion on slopes may be mapped by the symbols in Figure 2.4, and Figure 2.5 is a set of symbols from Nijman and Puigdefabregas (1978) for mapping detailed bar morphology. Slope mass-movement features and slope instability may be depicted by the symbols in Figure 2.6 after the Geological Society Engineering Group Working Party (1972). Figure 2.7 gives a set of symbols proposed by Blachut and Müller (1966) for mapping features associated with active glaciers. Symbols for mapping snow and ice *per se* are, however, given in Chapter 8.

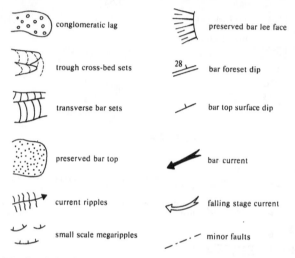

conglomeratic lag

trough cross-bed sets

transverse bar sets

preserved bar top

current ripples

small scale megaripples

preserved bar lee face

bar foreset dip

bar top surface dip

bar current

falling stage current

minor faults

Figure 2.5 Symbols for mapping detailed bar morphology (Nijman & Puigdefabregas 1978).

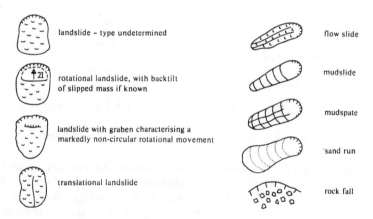

landslide – type undetermined

rotational landslide, with backtilt of slipped mass if known

landslide with graben characterising a markedly non-circular rotational movement

translational landslide

flow slide

mudslide

mudspate

sand run

rock fall

Figure 2.6 Symbols for detailed mapping of slope and mass-movement features and slope instability (Geological Society Engineering Group Working Party 1972).

From these elements symbols for complex types of landslide can be derived, for example:

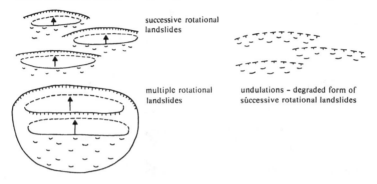

successive rotational
landslides

undulations – degraded form of
successive rotational landslides

multiple rotational
landslides

Where known, the category of material
involved in the slip should be shown
as follows:

    F   only fill involved

    D   only drift or superficial deposits
          involved

    S   only solid rocks involved

    F + D   fill and drift or superficial deposits
            involved

    D + S   drift or superficial deposits and
            solid rocks involved

These letters should be placed within the
landslide scar, for example:

 rotational landslide involving drift
or superficial deposits and solid rocks

 translational landslide involving
drift or superficial deposits only

The boundaries of all the above landslide symbols may be shown as certain, approximate or assumed.

It should be noted that the boundary in each case encloses all the area affected by the landslide,
whether by erosion or deposition.

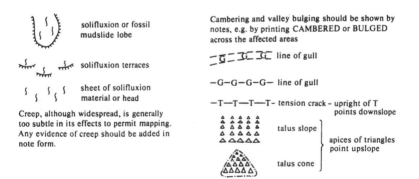

solifluxion or fossil
mudslide lobe

solifluxion terraces

sheet of solifluxion
material or head

Creep, although widespread, is generally
too subtle in its effects to permit mapping.
Any evidence of creep should be added in
note form.

Cambering and valley bulging should be shown by
notes, e.g. by printing CAMBERED or BULGED
across the affected areas

line of gull

–G–G–G–G– line of gull

–T—T—T—T- tension crack - upright of T
points downslope

talus slope

apices of triangles
point upslope

talus cone

Figure 2.6 – *continued.*

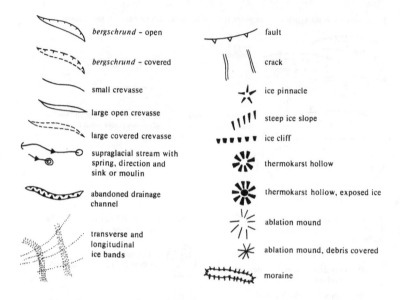

Figure 2.7 Symbols for mapping features associated with glaciers (Blachut & Müller 1966).

3 Slope profiling

This chapter consists of recommendations and procedures to be followed in producing a continuous slope profile (Sec. 3.1) and material necessary for particular techniques when carrying out slope survey single-handed or when measuring inaccessible slopes or cliffs (Sec. 3.2). Formulae for conversion between units of slope angle are also given in Table 3.2. For extensive tabulations of conversions between units of angular measurement see Young (1972), and for a review of profiling methods see Goudie (1981), pp. 62–5.

3.1 Continuous slope profiling

Brief recommendations for procedures to be followed in continuous slope profiling are given in this section. For a fuller description of techniques and equipment see Young *et al*. (1974), and for discussion concerning the merits of various criteria see Pitty (1966).

It is recommended that no measured length should be greater than 20 m or less than 2 m. The difference in angle between adjacent measured lengths should not exceed 2° on slopes less than 20°, or 4° on slopes greater than 20°, unless consecutive measured lengths are 2 m or less. These recommendations may, however, be varied in areas with high drainage densities or gentle slopes. Slope profiles should extend beyond divide and talweg until either a slope in the opposite direction to that surveyed has become clearly established (by a slope of 2° or more in the opposite direction) or a non-slope landform has been reached (e.g. flood plain). Near to crests of convex interfluves or centres of concave valleys the angle along crest or valley floor may exceed that perpendicular to its axis. In such cases the profile should be continued perpendicular to the crest or talweg, and the slope values should be identified as apparent slope values. A checklist of features to record during profiling is given in Table 3.1.

3.2 Particular methods of slope survey

Slope profiling normally requires two operators if it is to be carried out quickly and effectively by the majority of methods. Exceptions to this are

Table 3.1 Checklist of features to record during slope profiling.

ENTIRE PROFILE

Profile environment
(1) Identification, e.g. code number, date of survey, observer.
(2) Location, specified by (for example) co-ordinates from map grid.
(3) Geology. General features of area and detailed description of particular site, including superficial deposits. Source of data (e.g. field, map).
(4) Vegetation and land use of area, slope and profile line.
(5) Regolith and soil characteristics of area, slope and profile line.
(6) Microrelief of area, slope and profile line.
(7) Landforms. General features of area. Relation of profiled slope to nearby landforms and slopes. Relation of form along profile line to form on other parts of the slope on which profile is sited.
(8) River channel characteristics (if present). Width, depth, estimated speed and volume of flow. Whether slope is undercut.

Profile form
(1) Aspect at steepest point on the profile.
(2) Lateral slope at profile crest.
(3) Lateral slope at profile base.
(4) Plan curvature at steepest point on profile.

EACH MEASURED LENGTH
(1) Gradient – measured backwards and forwards.
(2) Ground surface distance.
(3) Vegetation or land use.
(4) Visible evidence of processes and materials, e.g. microrelief, evidence of mass movements, rock outcrops, stones, nature of regolith.
(5) Man-made features, e.g. hedges, which may be useful in locating the profile.
(6) Presence of disturbed ground.

Table 3.2 Conversions between units of angular measurement.

To convert to degrees from:		To convert from degrees to:	
radians	Deg = Rad. × 57.296	radians	Rad. = Deg/57.296
altan	Deg = $\tan^{-1}$ 0.001 (alog Alt/10)	altan	Alt. = 10 × log (1000 tan Deg)
% grade	Deg = $\tan^{-1}$ (%/100)	% grade	% = 100 tan Deg
gradient	Deg = $\tan^{-1}$ (1/Grd)	gradient	Grd = 1/tan Deg
feet per mile	Deg = $\tan^{-1}$ (ft/ml/5280)	feet per mile	ft/ml = 5280 tan Deg
metres per kilometre	Deg = $\tan^{-1}$ (m/km/1000)	metres per kilometre	m/km = 1000 tan Deg

the slope pantometer of Pitty (1968) and the two techniques for which details are given below. The Blong (1972) method uses the device illustrated in Figure 3.1, consisting of an upright on which is mounted a bar that can be made level and which is free to move vertically. It therefore records slope angles for short measured lengths (c. 1.0–1.5 m) in a way similar to the slope pantometer. Slope angle, θ, is given by

$$\tan \theta = \text{vertical height} \div \text{horizontal length},$$

and if the horizontal length is made to be 1.0 m, then

$$\tan \theta = \text{vertical height}.$$

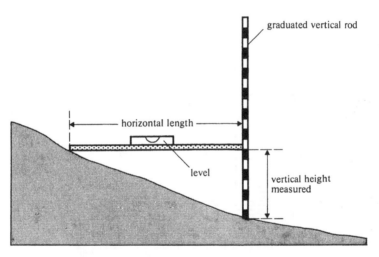

Figure 3.1 The Blong (1972) method of slope survey.

In the method of Gardiner and Dackombe (1977) angles are measured from the observer's eye to points marked on the ground surface along the profile. It therefore allows the use of measured lengths of any reasonable length. True slope angles for each measured length are given by use of the nomogram in Figure 3.2. A straight-edge is aligned from the origin to the point of intersection between the measured angle and the measured length, and the 'nomogram angle' is then read off from where this line intersects the appropriate operator eye height. The nomogram angle is then added to a measured inclination or is subtracted from a measured declination to give the true slope angle.

The following two techniques allow determination of slope gradient or height of inaccessible slopes. The first (Churchill 1979) necessitates the

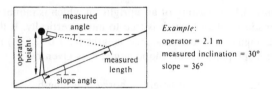

Example:
operator = 2.1 m
measured inclination = 30°
slope = 36°

Measured declination

$$\text{slope} = \frac{\text{measured}}{\text{angle}} - \frac{\text{nomogram}}{\text{angle}}$$

Measured inclination

$$\text{slope} = \frac{\text{measured}}{\text{angle}} + \frac{\text{nomogram}}{\text{angle}}$$

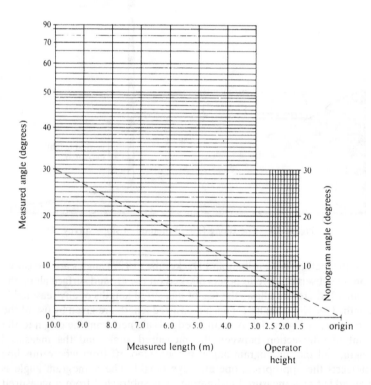

Figure 3.2 Nomogram for conversion of observed slope angle to actual slope angle in the method of survey of Gardiner and Dackombe (1977). The use of this is explained in the text.

beginning and end of the slope segments' being marked by features prominent enough to allow use of a range finder. The second method, for determination of heights of features such as cliffs, involves angular measurements only, but again requires that the point whose height is to be estimated can be identified from two points.

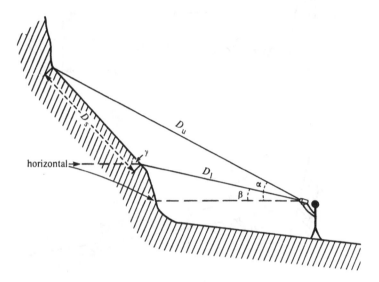

Figure 3.3 The Churchill (1979) method of slope survey.

The Churchill method is illustrated in Figure 3.3. The angles α and β are measured by theodolite or by Abney level. The distances D_u and D_l are measured by range finder. Slope segment length, D_s, and angle, γ, are then given by

$$D_s = [D_u^2 + D_l^2 - 2D_uD_l \cos (\alpha - \beta)]^{0.5},$$
$$\gamma = \cos^{-1}[(D_u^2 + D_s^2 - D_l^2)/2D_uD_s] + \alpha.$$

The second method is illustrated by Figure 3.4. This necessitates angular measurements from two points rather than use of a range finder. The angles α and β may be measured with either a theodolite or an Abney level. The separation of the two observations may be measured with a tape or chain. It is convenient to use a standardised distance for each set of determinations. The height above the plane on which the observer is

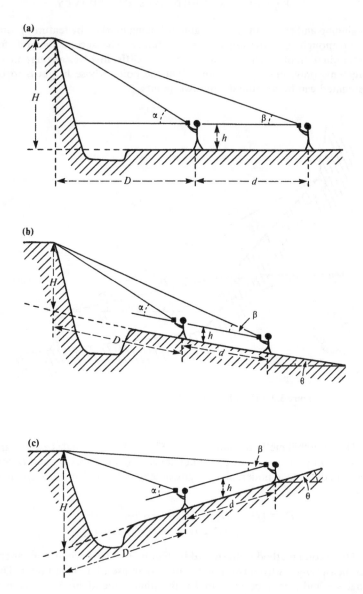

Figure 3.4 Determination of cliff heights: (a) from level plane; (b) from plane tilted away from feature; (c) from plane tilted towards feature.

standing (H) and the distance from the closest observation to the horizontal projection of the feature (D) are given by the following formulae.

(a) Sightings taken from a horizontal plane (Fig. 3.4a):

$$H = \{\tan \alpha \tan \beta/(\tan \alpha - \tan \beta)\} + h,$$
$$D = (H - h)/\tan \alpha.$$

(b) Sightings taken uphill on a slope of θ degrees (Fig. 3.4b):

$$H = h + [d \tan(\alpha - \theta)\tan(\beta - \theta)/\cos \theta\{\tan(\alpha - \theta) - \tan(\beta - \theta)\}],$$
$$D = \{(H - h)/\tan \alpha\} - (H - h)\sin \theta.$$

(c) Sightings taken downhill on a slope of θ degrees (Fig. 3.4c):

$$H = h + [d \tan(\alpha + \theta) \tan (\beta + \theta)/\cos \theta \{\tan(\alpha + \theta) - \tan(\beta + \theta)\}],$$
$$D = \{(H - h)/\tan \alpha\} + (H - h)\sin \theta.$$

Repeated readings taken at each station allow this method to be used for profiling inaccessible slopes.

4 Mapping landscape materials

This chapter is devoted to the mapping of solid and drift materials at various scales. It contains symbols and ornaments for both field mapping and the subsequent stage of presentation of cross sections and stratigraphic columns (Secs 4.1 & 2). Formulae and methods for the calculation of true and apparent dips and stratigraphic thicknesses are given in Sections 4.3 and 4.4. The emphasis of this chapter is on depicting the spatial pattern rather than on the detailed description of the materials themselves, which is the topic of Chapter 6. Although several symbols are given for peat, detailed work on peat-dominated sequences may require the use of a detailed scheme of peat symbols such as that given by Faegri and Gams (1937).

4.1 Geological mapping

Mapping at small scales (1 : 10 000 or smaller) involves essentially the mapping of structural attributes and gross rock or drift type. Symbols for geological structures are given in Figure 4.1 and should be recorded

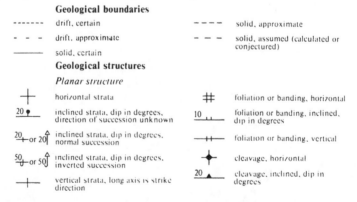

Figure 4.1 Symbols for small-scale geological mapping (Geological Society Engineering Group Working Party 1972, Compton 1962, Lahee 1961).

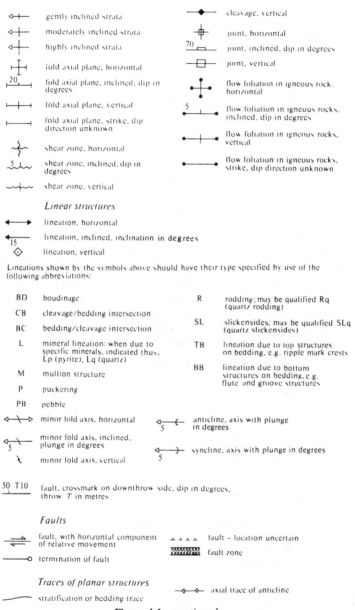

gently inclined strata

moderately inclined strata

highly inclined strata

fold axial plane, horizontal

20 fold axial plane, inclined, dip in degrees

fold axial plane, vertical

fold axial plane, strike, dip direction unknown

shear zone, horizontal

5 shear zone, inclined, dip in degrees

shear zone, vertical

cleavage, vertical

joint, horizontal

70 joint, inclined, dip in degrees

joint, vertical

flow foliation in igneous rock, horizontal

5 flow foliation in igneous rocks, inclined, dip in degrees

flow foliation in igneous rocks, vertical

flow foliation in igneous rocks, strike, dip direction unknown

Linear structures

lineation, horizontal

15 lineation, inclined, inclination in degrees

lineation, vertical

Lineations shown by the symbols above should have their type specified by use of the following abbreviations:

BD	boudinage	R	rodding; may be qualified Rq (quartz rodding)
CB	cleavage/bedding intersection		
BC	bedding/cleavage intersection	SL	slickensides; may be qualified SLq (quartz slickensides)
L	mineral lineation; when due to specific minerals, indicated thus, Lp (pyrite), Lq (quartz)	TB	lineation due to top structures on bedding, e.g. ripple mark crests
M	mullion structure	BB	lineation due to bottom structures on bedding, e.g. flute and groove structures
P	puckering		
PB	pebble		

minor fold axis, horizontal

5 minor fold axis, inclined, plunge in degrees

minor fold axis, vertical

5 anticline, axis with plunge in degrees

5 syncline, axis with plunge in degrees

50 T10 fault, crossmark on downthrow side, dip in degrees, throw T in metres

Faults

fault, with horizontal component of relative movement

termination of fault

fault - location uncertain

fault zone

Traces of planar structures

stratification or bedding trace

axial trace of anticline

Figure 4.1 – *continued.*

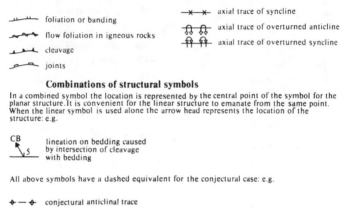

Combinations of structural symbols

In a combined symbol the location is represented by the central point of the symbol for the planar structure. It is convenient for the linear structure to emanate from the same point. When the linear symbol is used alone the arrow head represents the location of the structure: e.g.

CB   lineation on bedding caused
  5   by intersection of cleavage
        with bedding

All above symbols have a dashed equivalent for the conjectural case: e.g.

◆ — ◆   conjectural anticlinal trace

Figure 4.1 – *continued.*

directly on the field map along with a suitable abbreviated lithology name.

At larger scales, the maps or plans for both engineering and geomorphological purposes need to show more detail of the main rock mass characteristics, and the additional symbols given in Figures 4.2, 4.3 and 4.4 may be used as appropriate. Mapping in this context also includes the preparation of detailed elevations of natural or man-made cliffs and faces. Fuller details of the description of the main rock mass characteristics are given in Chapter 6. Figure 4.3 depicts a scheme of ornamentation recom-

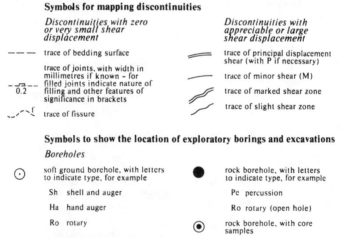

Figure 4.2 Additional symbols for large-scale geological mapping (Geological Society Engineering Group Working Party 1972).

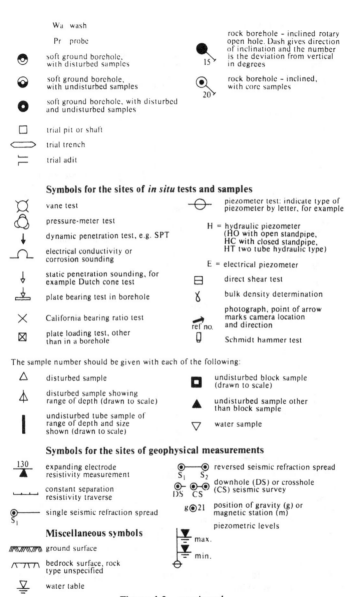

Wa wash

Pr probe

soft ground borehole, with disturbed samples

soft ground borehole, with undisturbed samples

soft ground borehole, with disturbed and undisturbed samples

trial pit or shaft

trial trench

trial adit

rock borehole - inclined rotary open hole. Dash gives direction of inclination and the number is the deviation from vertical in degrees

15

rock borehole - inclined, with core samples

20

Symbols for the sites of *in situ* tests and samples

vane test

pressure-meter test

dynamic penetration test, e.g. SPT

electrical conductivity or corrosion sounding

static penetration sounding, for example Dutch cone test

plate bearing test in borehole

California bearing ratio test

plate loading test, other than in a borehole

piezometer test: indicate type of piezometer by letter, for example

H = hydraulic piezometer (HO with open standpipe, HC with closed standpipe, HT two tube hydraulic type)

E = electrical piezometer

direct shear test

bulk density determination

photograph, point of arrow marks camera location and direction

ref no.

Schmidt hammer test

The sample number should be given with each of the following:

disturbed sample

disturbed sample showing range of depth (drawn to scale)

undisturbed tube sample of range of depth and size shown (drawn to scale)

undisturbed block sample (drawn to scale)

undisturbed sample other than block sample

water sample

Symbols for the sites of geophysical measurements

130

expanding electrode resistivity measurement

constant separation resistivity traverse

single seismic refraction spread

S_1

reversed seismic refraction spread

S_1 S_2

downhole (DS) or crosshole (CS) seismic survey

DS CS

position of gravity (g) or magnetic station (m)

g⊙21

piezometric levels

max.

min.

Miscellaneous symbols

ground surface

bedrock surface, rock type unspecified

water table

Figure 4.2 – *continued.*

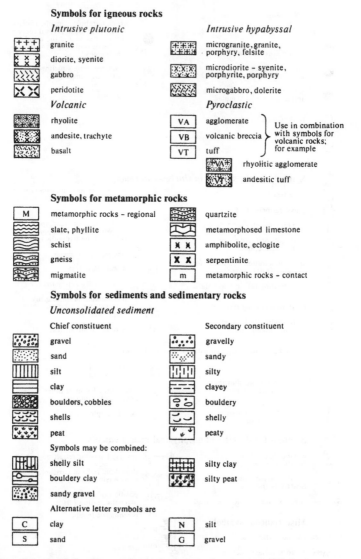

Symbols for igneous rocks

Intrusive plutonic

granite

diorite, syenite

gabbro

peridotite

Volcanic

rhyolite

andesite, trachyte

basalt

Intrusive hypabyssal

microgranite, granite, porphyry, felsite

microdiorite – syenite, porphyrite, porphyry

microgabbro, dolerite

Pyroclastic

VA agglomerate

VB volcanic breccia

VT tuff

Use in combination with symbols for volcanic rocks; for example

rhyolitic agglomerate

andesitic tuff

Symbols for metamorphic rocks

M   metamorphic rocks – regional

slate, phyllite

schist

gneiss

migmatite

quartzite

metamorphosed limestone

X X   amphibolite, eclogite

X X   serpentinite

m   metamorphic rocks – contact

Symbols for sediments and sedimentary rocks

Unconsolidated sediment

Chief constituent

gravel

sand

silt

clay

boulders, cobbles

shells

peat

Secondary constituent

gravelly

sandy

silty

clayey

bouldery

shelly

peaty

Symbols may be combined:

shelly silt

bouldery clay

sandy gravel

silty clay

silty peat

Alternative letter symbols are

C   clay

S   sand

N   silt

G   gravel

Figure 4.3 Symbols for rock type and hydrogeology (Geological Society Engineering Group Working Party 1972).

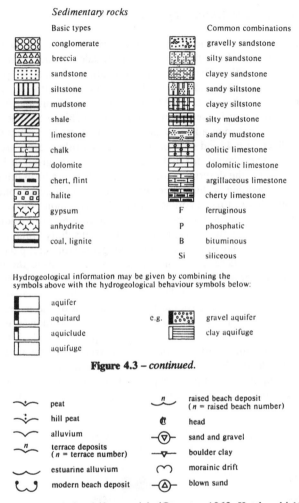

Sedimentary rocks

Basic types		Common combinations	
	conglomerate		gravelly sandstone
	breccia		silty sandstone
	sandstone		clayey sandstone
	siltstone		sandy siltstone
	mudstone		clayey siltstone
	shale		silty mudstone
	limestone		sandy mudstone
	chalk		oolitic limestone
	dolomite		dolomitic limestone
	chert, flint		argillaceous limestone
	halite		cherty limestone
	gypsum	F	ferruginous
	anhydrite	P	phosphatic
	coal, lignite	B	bituminous
		Si	siliceous

Hydrogeological information may be given by combining the symbols above with the hydrogeological behaviour symbols below:

	aquifer			
	aquitard	e.g.		gravel aquifer
	aquiclude			clay aquifuge
	aquifuge			

Figure 4.3 – *continued.*

	peat		raised beach deposit (*n* = raised beach number)
	hill peat		head
	alluvium		sand and gravel
	terrace deposits (*n* = terrace number)		boulder clay
	estuarine alluvium		morainic drift
	modern beach deposit		blown sand

Figure 4.4 Symbols for drift materials (Compton 1962, Kottlowski 1965).

mended by the Geological Society Engineering Group Working Party (1972) specifically for use in the preparation of maps and plans depicting rock mass character. These ornaments may be combined to describe lithology fully and may be used in conjunction with the structural symbols of Figures 4.1 and 4.2 and the symbols for the location of site investigations shown in Figure 4.2.

4.2 Logging sedimentary sequences

This section contains an integrated set of symbols for logging sedimentary sequences where the ultimate aim is a detailed palaeo-environmental reconstruction rather than the characterisation of material properties. This set of symbols should be used in conjunction with the section on sediment description (Sec. 6.5). Ornament for each unit may be built up using the ornament for sediment type as a base (Fig. 4.5). The nature of the boundary between adjacent units may be described using the terms given in Section 6.4, and symbols for these and other terms are given in Figure 4.6. The symbols for bedding-plane structures given in Figure 4.6 may be used for the description of bedding within a uniform lithology, for the type of bedding within heterolithic units in a sequence, and as supplementary descriptors of contact nature. It is common practice to make the width of the unit in the final stratigraphic column proportional to the mean grain size or the percentage of sand. Subsidiary columns may be added for the display of information on biogenic structure, palaeocurrent directions, etc., wherever this information would confuse the basic pattern. Symbols for palaeocurrent information are given in Figure 4.7. The symbols should be aligned to show the direction of flow relative to north, which in a section or log should be from the top of the page. Where a number of measurements are made, the standard deviation sector and the number of readings should be added to the upcurrent end of the symbol. In the case of alignment symbols such as striae, two symmetric sectors should be added. Current strength may be shown by the length of the symbol if required. Symbols to depict sedimentary structures are given in Figure 4.8 and should normally be used as an overlay on the lithology

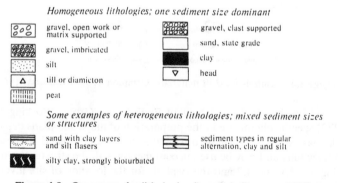

Homogeneous lithologies; one sediment size dominant

gravel, open work or matrix supported

gravel, imbricated

silt

till or diamicton

peat

gravel, clast supported

sand, state grade

clay

head

Some examples of heterogeneous lithologies; mixed sediment sizes or structures

sand with clay layers and silt flasers

silty clay, strongly bioturbated

sediment types in regular alternation, clay and silt

Figure 4.5 Ornaments for lithologies (based on Boersma 1975).

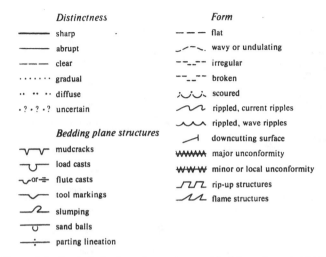

Figure 4.6 Symbols for boundary nature and bedding plane structures.

ornament. Smaller symbols, which may be used as alternatives or where the feature is of subsidiary importance or obscures other information, are given in brackets. Finer detail of flaser, wavy and lenticular bedding may be represented using the symbolism shown in Figure 6.16. The range of cross-bedding types covered by Allen's (1963) classification may be displayed by combining the appropriate symbol from Figure 4.8 with the

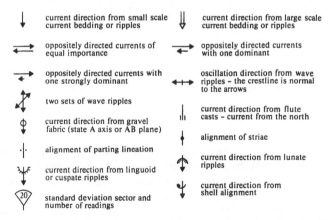

Figure 4.7 Symbols for palaeocurrent information (based on Boersma 1975).

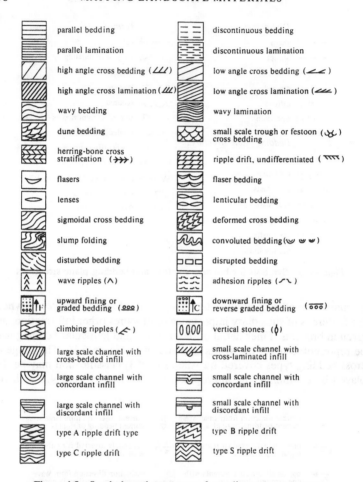

Figure 4.8 Symbols and ornaments for sedimentary structures.

Greek letter notation. The classification of ripple bedding of Jopling and Walker (1967) may be similarly represented or the symbols shown in Figure 4.8 may be used. Symbols for biogenic structures, fossils and miscellaneous features are shown in Figure 4.9 and an example log using the scheme is shown in Figure 4.10. Symbols from Figure 4.2 may be used to show the location of samples, photographs, etc.

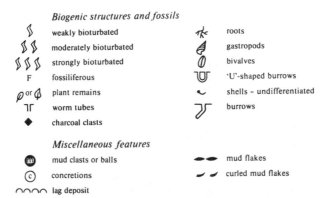

Biogenic structures and fossils

	weakly bioturbated		roots
	moderately bioturbated		gastropods
	strongly bioturbated		bivalves
F	fossiliferous		'U'-shaped burrows
or	plant remains		shells – undifferentiated
	worm tubes		burrows
◆	charcoal clasts		

Miscellaneous features

	mud clasts or balls		mud flakes
c	concretions		curled mud flakes
	lag deposit		

Figure 4.9 Symbols for biogenic structures, fossils and miscellaneous features (compiled from Boersma 1975, Cherven 1978, Taylor & Woodyer 1978).

4.3   Conversion between true and apparent dips

The recovery of true dip from measured apparent dips is a common operation and three alternative methods are given. The use of a stereonet is familiar to geologists and requires a printed net as a base. This is not provided here because of the small page size. Both the trigonometric and graphical methods require a minimum of equipment and may be used in the field. The construction of geological cross sections requires that true dip be converted into an apparent dip along the line of the section, and a simple trigonometric method for doing this is given.

Stereographic projection method

Plot the two apparent dips by rotating the film over the stereonet until the azimuth of the dip is aligned along the east–west axis. The amount of dip is plotted inwards from the circumference by the appropriate number of gradations. Rotate the film until both apparent dips lie on the same great circle. The amount of true dip may then be read directly from the intersection of the great circle and the east–west axis. Mark the point of intersection and rotate the film back to the reference position. The azimuth of the true dip may now be read.

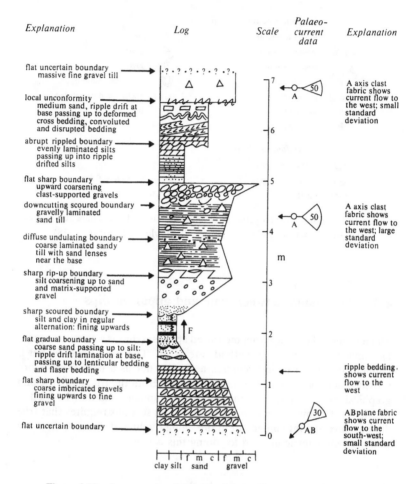

Figure 4.10   An example of the use of the sediment logging symbols.

Trigonometric method

A = azimuth of the first apparent dip; B = azimuth of the second apparent dip; a = amount of the first apparent dip; b = amount of the second apparent dip.

Let $K = A - B$ and let

$$\tan X = \text{cosec}\, K\, \{(\cot a \tan b) - \cos K\}\,.$$

The true dip direction (C) and the amount of true dip (c) are given by

$$C = A - \arctan X,$$
$$c = \arctan (\tan a/\cos C).$$

In the special case where A and B are normal to each other,

$$\tan X = \cot a \tan b,$$
$$C = A - \arctan X,$$
$$c = \arctan (\tan a/\cos c).$$

Graphical method

Draw a circle of convenient diameter and insert the north–south and east–west axes (Fig. 4.11). Draw in the lines AB and AC at the azimuths of the two apparent dips. From the centre of the circle at A draw perpendiculars (AX and AY) to the lines AB and AC. Construct the triangle ABX making the angle ABX the value of the first apparent dip b. Repeat for the triangle ACY making the angle ACY the second apparent dip c. Draw the line CB and the perpendicular to it, AD. Construct the triangle AZD. The angle AZD is the amount of true dip, d.

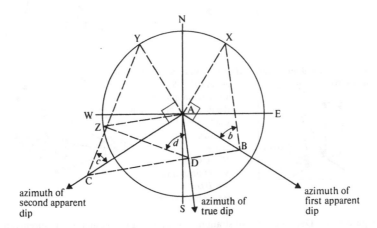

Figure 4.11 A graphical method for the conversion of apparent to true dip.

(a) Slope and dip in opposite directions, their sum being greater than 90°

$T = H \sin d + V \cos d$
or
$T = S \sin(s + d)$

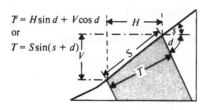

(b) Slope and dip in opposite directions, their sum being less than 90°

$T = H \sin d + V \cos d$
or
$T = S \sin(s + d)$

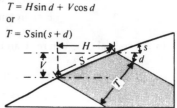

(c) Slope and dip in the same direction, with dip greater than slope

$T = H \sin d - V \cos d$

$T = S \sin(d - s)$

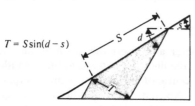

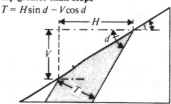

(d) Slope and dip in the same direction, with dip less than slope

$T = V \cos d - H \sin d$

$T = S \sin(s - d)$

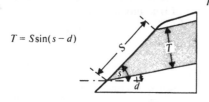

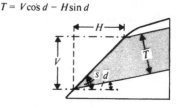

(e) Inclined bed measured on a vertical surface

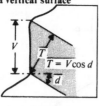

$T = V \cos d$

(f) All the solutions given above assume that S and H are measured parallel to the direction of true dip. Corrections for oblique measurements are given below.

$H = Q \cos b$
$S = D \sin c / \sin s$
$S = D \cos a$
$\tan s = \tan c / \cos b$

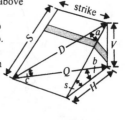

Figure 4.12 Determination of stratigraphic thickness from outcrop measurements (after Kottlowski 1965, Compton 1962).

True dip to apparent dip along a line of section

Using the notation shown in Figure 4.12f, apparent dip along the line of section may be obtained from

$$\tan c = \tan s \cos b \, ,$$

where c is the apparent dip, b is the angle of divergence of the line of section from the direction of true dip, s is the angle of true dip.

4.4 Calculation of stratigraphic thickness

Methods of calculating the stratigraphic thickness of members from various field measurements are shown in Figure 4.12. For other geometric problems see Chapter 13, where a variety of trigonometric formulae are given.

5 Geophysical methods of subsurface investigation

This chapter is concerned with the two main geophysical methods of subsurface investigation used by geomorphologists. The common problem of 'How thick is this drift and what is its physical nature in depth?' may be answered under favourable circumstances by use of either the seismic refraction method (dealt with in the first half of this chapter) or by the resistivity method (the subject of the second half). Quite often the two methods used together may give good results and the reader is referred to the excellent.discussion of the pros and cons of each method by Griffiths and King (1965).

5.1 The seismic refraction method

This section provides the information and materials for field use of the seismic refraction method by an operator who is already familiar with the method. It is in no sense an adequate introduction to the method; the user who requires such an introduction is referred to one of the excellent texts by Griffiths and King (1965), Telford et al. (1976), Parasnis (1979) or Redpath (1973). The assumption is made throughout that only single channel equipment with either a sledgehammer or falling-weight shock source is available. Ideally the enhancement type of seismograph with full wavetrain display should be used in preference to those that merely display the first arrival time on a decade counter. The user should be fully aware of any limitations placed on the reliability of his results by the equipment used and should be thoroughly familiar with the manufacturer's instructions. No guidance on the possible ranges of depth penetration or maximum line lengths is given since they are strongly dependent on local conditions and the equipment used. Reliance on the enhancement facility of some machines can lead to the use of excessively long lines and the use of relatively poorly defined breaks. To avoid misleading results only clear, well defined breaks should be recorded. Profiling methods are not covered here but details of them may be found in standard texts such as Dobrin (1976). If profiling methods are used, care should be taken that small variations in velocities, especially surface material velocity, and inaccuracies in measured lengths and surface topographic effects are not interpreted as variations in the depth of the refractor. As a general rule all seismic lines should be shot in both directions to avoid some of the possible misinterpretations explained in Section 5.2. Although seismic reflection methods may also be used for geomorphological work, the equipment is not as generally available and they are therefore not covered here.

Section 5.2 contains a brief description of some of the more commonly encountered results as a guide to the interpretation of time–distance graphs, and Section 5.3 contains nomograms for the rapid solution of seismic equations in the field. A full range of formulae are given in Sections 5.4 and 5.5 and a simple method of determining the depth of overburden is to be found in Section 5.6. Table 5.2 gives typical values of P-wave velocity for various rocks, sediments and fluids.

5.2 Time–distance plots for various simple geological structures

The graphs shown in Figure 5.1 show the plots of first arrival times against distance for a variety of simple subsurface situations. Only where they are necessary to define a structure are the plots of second arrival time shown. Figure 5.1a shows the simplest of all situations, that of an extensive interface between a lower high-velocity formation and an upper low-velocity material with the interface being parallel to the ground surface. In this case the velocity of the upper layer (V_0) and that of the lower layer (V_1) may be determined directly from the gradients of the graph segments. In the case of the situation shown in Figure 5.1b, where the interface dips relative to the ground surface, the true velocity of the lower layer can only be obtained by carrying out line reversal. That entails expanding the line from a detector at one end of the line and then moving the detector to the last occupied shot position at the far end of the line and working towards the first detector position. Note that, where the velocity contrast between the two layers is low, the line shot in the downdip direction may not show a clear two segment pattern and the lower layer will not be detected unless the downdip apparent velocity (V_{1d}) is significantly greater than the surface material velocity (V_0).

Figure 5.1c and d shows the likely results of shooting in the vicinity of a buried cliff. The former shows the situation when the line is shot in the off-cliff direction and can be mistaken for the situation shown in Figure 5.1g unless the line is reversed. The reversed situation (Fig. 5.1d) can itself be mistaken for the simple situation shown in Figure 5.1a and will

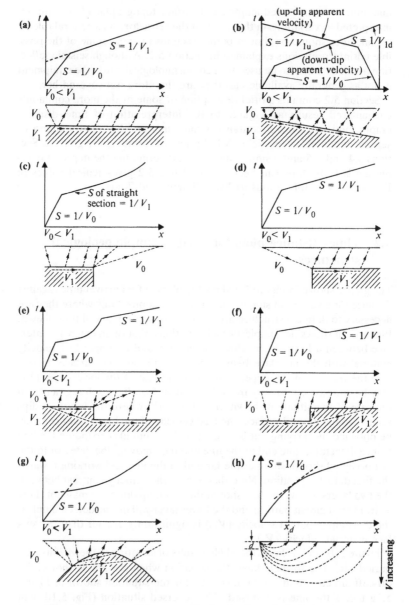

Figure 5.1 Time–distance graphs for simple geological situations.

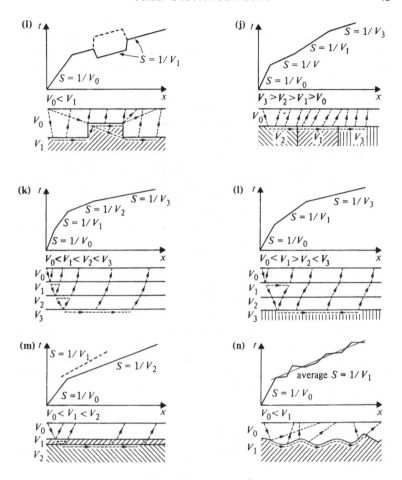

Figure 5.1 – *continued.*

give an excessive value for the depth of the interface. Obviously it is essential to shoot all lines in both directions in order to be reasonably sure of the correct interpretation.

Figure 5.1e and f shows examples of the plots obtained when shooting across a faulted area or a small step in the high-velocity lower layer. In cases where the displacement is small the central curved section of the plot may not be clearly defined. Both of these two examples are clear extensions of the cases shown in the Figure 5.1c and d.

Figure 5.1g shows the effect of an isolated dome-shaped mass of high-velocity material and Figure 5.1h illustrates the plots given when there is a steady downward increase in the velocity of the material instead of a stepwise increase. This latter case could be expected to occur where unconsolidated sediments overlie deeply weathered bedrock.

Figure 5.1i shows the plot given by a transect across a topographic high in the lower layer. The dotted line shows the reverse situation, that of a transect across a topographic low or trough. Figure 5.1j shows the effect of lateral variations in the velocity of the lower high-velocity layer.

Figure 5.1k shows the case in which multiple parallel layers show a successive increase in velocity with depth. The case shown in Figure 5.1b can be extended to multiple dipping layers in which the slope of each line segment represents the apparent velocity of each successive layer. Care must be taken when interpreting plots such as that in Figure 5.1k to ensure that as far as is possible the case shown in Figure 5.1l is recognised. In this case the third layer has a velocity that is lower than that of the layer immediately above it and it cannot be detected by refraction methods. The only means of detecting such layers is to shoot a down-hole survey in which the velocity variation with depth is investigated.

A similar pitfall with the refraction method is the failure to recognise the situation shown in Figure 5.1m where the second layer is too thin in relation to its depth to give first arrivals. This case can be recognised if the second arrivals, shown dashed on the figure, are plotted. An estimate of the maximum thickness of any such suspected layer can be made using the method of Hawkins and Maggs (1961). Finally, the common situation of a scatter of points instead of an ideal straight line may be the product of the case shown in Figure 5.1n of an irregular interface, or of velocity variations in either or both of the materials, or errors or lack of precision in timing and distance measurement. When such conditions appear to occur it is wise to shoot a number of short lines within the main spread in order to get good surface material velocity control.

5.3   Nomograms for seismic calculations

The three nomograms included in this section may be used for preliminary analysis of data in the field at the time of shooting. A restriction is that all of the nomograms (based on Meidav 1960) are valid for only a single interface which is parallel or closely subparallel to the ground surface. All of the nomograms are scaled appropriately for use with SI units; although they can be used with Imperial units the scales may be inconvenient.

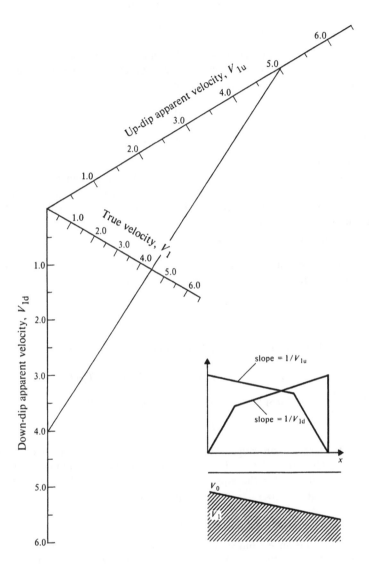

Figure 5.2 True velocity nomogram (based on Meidav 1960).

True velocity nomogram

The true velocity nomogram (Fig. 5.2) can be used to determine the true velocity, V_1, of a refractor from the two apparent velocities, V_{1u} and V_{1d} (Fig. 5.2, inset). It solves the equation

$$V_1 = 2 \cos \alpha (1/V_{1u} + 1/V_{1d}) ,$$

where the angle of dip (α) is small and $\cos \alpha \approx 1.0$.

Method of use
Connect values of V_{1u} and V_{1d} and read off V_1 on the centre scale; e.g. $V_{1d} = 4.0$, $V_{1u} = 5.0$ gives $V_1 \approx 4.5$. Any convenient units may be used. The apparent velocities may be halved to bring them on scale and the result must then be doubled.

Intercept time nomogram

The intercept time nomogram (Fig. 5.3) can be used to solve seismic refraction equations which contain the common factor of

$$A = V_1 V_0 / (V_1^{2} - V_0^2)^{1/2}.$$

It may be used to compute the following:

(a) The depth to the refracting horizon (Fig. 5.3, inset A),

$$Z_0 = T_i V_1 V_0 / 2 (V_1^2 - V_0^2)^{1/2} = T_i A / 2.$$

(b) The vertical displacement of a fault (Fig. 5.3, inset B),

$$\Delta Z = \Delta T V_1 V_0 / (V_1^2 - V_0^2)^{1/2}.$$

(c) The depths to an irregular or slightly dipping interface at each detector. The value of A derived from the intercept time nomogram may be used to solve the equation repeatedly for the unknown depths. This equation is valid for low angles of dip only.

$$Z_j = A(T_j - X_j/V_1) - Z_1,$$

where X_j = distance from shot to jth detector, Z_1 = known depth at the shot point, Z_j = depth to interface at jth detector, T_j = total time to the jth detector.

Method of use
Step 1: Connect V_1 to V_0; read off A.
Step 2: Connect A to T_i or ΔT; read off Z or ΔZ.

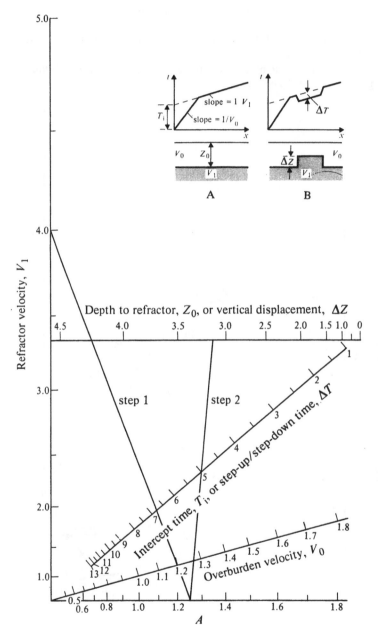

Figure 5.3 Intercept time nomogram (based on Meidav 1960).

For example $V_1 = 4.0$ to $V_0 = 1.2$ gives $A = 1.25$; then $A = 1.25$ to $T_i = 5.0$ gives $Z = 3.2$.

Using V_1 and V_0 in metres per millisecond and T_i and ΔT in milliseconds gives values of Z and ΔZ in metres.

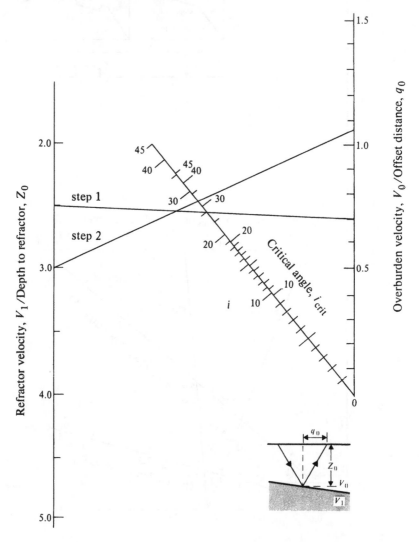

Figure 5.4 Critical angle and offset distance nomogram (based on Meidav 1960).

Critical angle and offset distance nomogram

The critical angle and offset distance nomogram (Fig. 5.4) may be used to determine

(a) the critical angle of incidence, i_{crit}, from $i_{crit} = \sin^{-1}(V_0/V_1)$;

(b) the offset distance q_0 (Fig. 5.4, inset), using $q_0 = Z_0 \tan i_{crit}$.

Method of use
Step 1: V_1 and V_0 and read off i_{crit} .
Step 2: Transfer the value of i_{crit} from Step 1 to the i scale.
        Connect i and z_0 and read off q_0.

For example, $V_0 = 0.7$ and $V_1 = 2.5$ gives $i_{crit} = 28°$, then 'i' $= i_{crit} = 28°$, $Z_0 = 2.5$ gives $q_0 = 1.06$.

V_1 and V_0 should both be in degrees. q_0 is given in the same units as those used for Z_0.

5.4 Seismic formulae

Seismic refraction formulae for multiple dipping layers are presented in Table 5.1 and characteristic wave paths, time–distance graphs and notation are shown in Figure 5.5. The formulae are based upon those of Ewing *et al.* (1939) and are presented here in an operational sequence, computation proceeding from left to right along each successive row for each successive layer.

It is important to note that dips of successively deeper layers (ω) are given with respect to the layers above them, not with respect to horizontal. Layer thickness (Z) is measured normal to the lower boundary, and the formulae given can be used at both updip and downdip ends of the transect using the appropriate intercept times (T_{iu}, T_{id}). For situations with more than four layers it is only necessary to increase the appropriate subscripts, although the formula for layer thickness acquires an extra term for each additional layer. These formulae can be collapsed to yield solutions for multiple parallel layers.

5.5 Corrections used in refraction analysis

Elevation correction

The correction to delay time at the shot point which should be subtracted from the observed delay time is

$$C = (e - h - d)(V_1^2 - V_0^2)^{1/2}/V_1 V_0$$
$$= (e - h - d)\cos i_{crit}/V_0,$$

Table 5.1 Seismic refraction formulae for multiple dipping layers cases. For notation see Figure 5.5 (based on Ewing *et al.* 1939). Computation should proceed from left to right for each successive layer.

Layer	$\alpha_{0,n}$ = angle of incidence at the first interface of the critical ray to the nth interface (downdip)	$\beta_{0,n}$ = angle of incidence at the first interface of the critical ray to the nth interface (updip)	$i_{\text{crit}_{n-1,n}}$ = the critical angle of incidence at the nth interface
0			
1	$\alpha_{0,2} = \sin^{-1}(V_0/V_{2u}) + \omega_{0,1}$	$\beta_{0,2} = \sin^{-1}(V_0/V_{2d}) - \omega_{0,1}$	$i_{\text{crit}_{0,1}} = 0.5\{\sin^{-1}(V_0/V_{1d}) + \sin^{-1}(V_0/V_{1u})\}$
2	$\alpha_{0,3} = \sin^{-1}(V_0/V_{3u}) + \omega_{0,1}$	$\beta_{0,3} = \sin^{-1}(V_0/V_{3d}) - \omega_{0,1}$	$i_{\text{crit}_{1,2}} = 0.5\{\sin^{-1}(V_1 \sin \alpha_{0,2}/V_0) + \sin^{-1}(V_1 \sin \beta_{0,2}/V_0)\}$
n	$\alpha_{0,n} = \sin^{-1}(V_0/V_{(n+1)u}) + \omega_{0,1}$	$\beta_{0,n} = \sin^{-1}(V_0/V_{(n+1)d}) - \omega_{0,1}$	$i_{\text{crit}_{2,3}} = 0.5\{\sin^{-1}(V_2 \sin \alpha_{1,3}/V_1) + \sin^{-1}(V_2 \sin \beta_{1,3}/V_1)\}$
			$i_{\text{crit}_{n,n+1}} = 0.5\{\sin^{-1}(V_n \sin \alpha_{n-1,n+1}/V_{n-1}) + \sin^{-1}(V_n \sin \beta_{n-1,n+1}/V_{n-1})\}$

Layer	$\omega_{n-1,n}$ = the angle of dip of the nth interface with respect to the one above it	V_{n+1} = velocity of the (n + 1)th layer	Z_n = the thickness of the nth layer measured at right angles to the lower bounding interface
0	$\omega_{0,1} = 0.5\{\sin^{-1}(V_0/V_{1d}) - \sin^{-1}(V_0/V_{1u})\}$	$V_1 = V_0/\sin i_{\text{crit}_{0,1}}$	$Z_0 = V_0(T_{i_1}/2 \cos i_{\text{crit}_{0,1}})$
	$\omega_{1,2} = 0.5\{\sin^{-1}(V_1 \sin \alpha_{0,2}/V_0) - \sin^{-1}(V_1 \sin \beta_{0,2}/V_0)\}$	$V_2 = V_1/\sin i_{\text{crit}_{1,2}}$	$Z_1 = \dfrac{V_1\{T_{i_2} - Z_0(\cos \alpha_{0,2} + \cos \beta_{0,2})/V_0\}}{2 \cos i_{\text{crit}_{1,2}}}$
	$\omega_{2,3} = 0.5\{\sin^{-1}(V_2 \sin \alpha_{1,3}/V_1) - \sin^{-1}(V_2 \sin \beta_{1,3}/V_1)\}$	$V_3 = V_2/\sin i_{\text{crit}_{2,3}}$	$Z_2 = \dfrac{V_2\{T_{i_3} - Z_1(\cos \alpha_{1,3} + \cos \beta_{1,3})/V_1 - Z_0(\cos \alpha_{0,3} + \cos \beta_{0,3})/V_0\}}{2 \cos i_{\text{crit}_{2,3}}}$
	$\omega_{n,n+1} = 0.5\{\sin^{-1}(V_n \sin \beta_{n-1,n+1}/V_{n-1}) - \sin^{-1}(V_n \sin \alpha_{n-1,n+1}/V_{n-1})\}$	$V_n = V_{n-1}/\sin i_{\text{crit}_{n-1,n}}$	$Z_n = \dfrac{V_n\left[T_{i_{n+1}} - \left\{\displaystyle\sum_{i=0}^{i=n-1} Z_i(\cos \alpha_{i,n+1} + \cos \beta_{i,n+1})/V_i\right\}\right]}{2 \cos i_{\text{crit}_{n,n+1}}}$

(a)

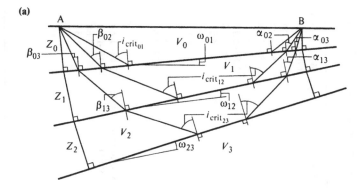

(b)

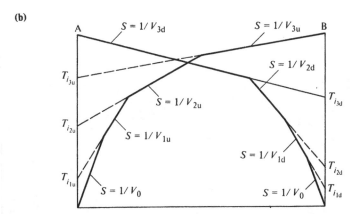

Figure 5.5 Characteristic wavepaths (a) and time–distance graph (b) for multiple dipping layers. Note that the angle of dip of each interface (ω) is given relative to that of the next higher interface (based on Ewing *et al.* 1939).

where e = elevation at shot point, h = depth of shot hole, d = elevation of datum plane.

A similar correction is applied at the detector using the same equation but omitting the shot hole depth term. The total elevation correction to be subtracted from the intercept time is the sum of these two corrections.

Table 5.2 Velocities of primary seismic waves in various materials. Compiled from various sources including Lahee (1961), Jakosky (1957), Redpath (1973), Dobrin (1976) and Kesel (1976).

Sediments and weathered rocks	Wave velocity ($m\ ms^{-1}$)
light, dry top soil	0.18–0.28
moist loamy or silty soil	0.30–0.40
wet loam	c. 0.76
dense wet clay	0.90–1.80
rubble or gravel	0.60–0.80
sandy clay	0.98–1.16
wet sand } large variations in P wave	0.43–1.40
sand } velocity are reported by different authors	1.40–2.56
till	1.70–2.26
head	1.60–2.10
moraine	0.76–1.53
wet moraine	1.53–2.13
loose rock talus	0.38–0.76
cemented sand	0.85–0.98
cemented sandy clay	1.16–1.28
cemented agglomerate	1.53–1.83
weathered head	0.40–0.60
weathered shale	1.22–1.98
weathered sandstone	c. 1.44
granite – highly decomposed	0.47
granite – badly fractured, part decomposed	0.67
granite – softened, part decomposed	3.20
basalt – weathered and fractured	2.74–4.27

Unweathered rocks	Wave velocity ($m\ ms^{-1}$)
shale	2.74–4.27
sandstone	1.83–4.27
chalk	1.83–3.96
limestone	2.13–6.10
salt	1.83–5.18
granite	3.96–6.10
granodiorite	4.57
basalt	5.58
dolerite	6.00
gabbro	3.38
phylite	3.05–3.35
slate	2.29–4.69
Fluids	
water } vary with temperature and salinity	1.43–1.68
sea water }	1.46–1.53
ice	3.68
oil	1.31

Combined elevation and weathering correction

$$C = (e_s + e_d - h - t - 2d)(V_2^2 - V_1^2)^{1/2}/V_1 V_2$$
$$+ t(V_2^2 - V_0^2)^{1/2}/V_2 V_0,$$

where e_s = shot point elevation, e_d = detector elevation, t = thickness of weathered layer, V_0 = velocity of weathered layer, V_1, V_2 = first and second layer velocities. C is subtracted from the intercept time.

5.6 A rapid method for determining the depth to bedrock beneath superficial deposits

This simple and rapid method may be used where the contrast between bedrock and drift velocities is large, with bedrock velocity being at least three times that of the drift cover. The overburden should be reasonably homogeneous with respect to velocity, depth variation should be smooth, and the angle of dip of the interface small. Using the notation shown in Figure 5.6,

$$\text{depth to bedrock } (Z_c) = \tfrac{1}{2}V_0(T_{AC} + T_{BC} - T_{AB}).$$

This method is a variant of the plus–minus method of Hagedoorn (1959), in which the term $V_1 V_0/(V_1^2 - V_0^2)^{1/2}$ is close to unity.

T_{AC}, T_{BC} and T_{AB} are the travel times over the three distances AC, BC and AB respectively. A short conventional line should be shot to establish the velocity of the drift material.

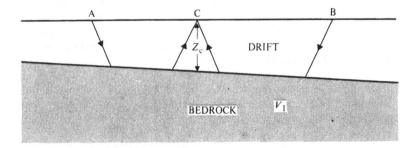

Figure 5.6 Layout and notation for a simple method of establishing the depth of bedrock.

5.7   The resistivity method

This section provides some information and materials for the field use of simple resistivity methods by an operator who is already familiar with the methods. The user who requires a complete introduction to the theory and practice of resistivity surveying should refer to one of the standard introductory geophysics texts such as Griffiths and King (1965), Telford *et al.* (1976) or Parasnis (1979). Bhattacharya and Patra (1968), Keller and Frischknecht (1966) and Van Nostrand and Cook (1966) are useful advanced texts that cover the interpretation of resistivity data.

For most purposes it is preferable to use low-frequency (60 Hz to *c.* 5 Hz) alternating current equipment. This obviates the need to use non-polarising electrodes or to measure and allow for spontaneous potentials. The resistivity measured with alternating current equipment will normally be rather lower than the true direct current resistivity, but the advantage of speedy operation usually outweighs this disadvantage. Where a direct current source is used, porous-pot potential electrodes must be used and the spontaneous potentials measured before the current is applied. The spontaneous potential may then be subtracted, either directly or by the application of a compensating voltage, from the potential measured when the current is flowing.

Although alternating current sources eliminate the spontaneous potential effect and give a better signal to noise ratio, they may give erratic readings. This may be due either to inductive coupling between adjacent long current and potential leads or to current leakage under wet ground conditions.

The choice and placing of electrodes is important. Non-polarising electrodes must be used with direct current sources. Contact resistance between the electrode and the ground should be kept to a minimum by ensuring that the electrodes are firmly embedded. Contact may be improved in dry ground by watering the electrodes or by the use of multiple stake or pot electrodes. On bare rock a wad of moist clay may be used to attach the electrode.

Section 5.8 deals with electrode configurations and the choice of an appropriate one for the work in hand while Section 5.9 deals with simple methods of interpreting two-layer cases. Space does not permit a full treatment of three- or multilayer interpretation procedures, but references are given in Section 5.9 to sources for this information.

Section 5.10 deals briefly with resistivity profiling and a simple method of overburden survey is given in Section 5.11. Section 5.12 deals with the electrical properties of common Earth materials.

5.8 Electrode configurations

The Wenner array

The Wenner array (Fig. 5.7a) is symmetrical with a uniform electrode spacing. For vertical electrical sounding (resistivity drilling) the whole array is expanded about the centre point (Fig. 5.7b) and its use for horizontal profiling involves moving the whole array along in successive steps (Fig. 5.7c). The apparent resistivity measured with the Wenner array is given by

$$\rho_a = K\Delta V/I \text{ ohm metres,}$$

where ΔV = the potential difference across the inner pair of electrodes (in volts or millivolts), I = the current applied across the outer pair of electrodes (in amperes or milliamperes), K = a term with the dimension of length, the magnitude of which depends on the geometry of the array. In

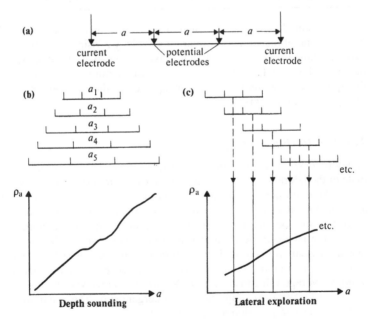

Figure 5.7 The Wenner array and its use for depth sounding and lateral exploration.

the Wenner array $K = 2\pi a$, where a is the electrode spacing in metres.

$\Delta V/I$ is usually read directly from the instrument as the total resistance. K takes on values dependent on the electrode spacing when the array is used for electric drilling but remains constant when horizontal profiling.

Lateral variations in resistivity may be recognised and distinguished from vertical variations by using the electrode switching method proposed by Carpenter (1955). In this method the geometric arrangement of the Wenner array is retained but three measures of apparent resistivity are made using the electrodes connected in different ways. The electrode arrangements, measured resistances and apparent resistivities are given in Table 5.3. If the three apparent resistivities all follow the same trend, it

Table 5.3 Electrode arrangements, measured resistances and apparent resistivities used in the tri-potential method of Carpenter (1955).

Electrode arrangement	Resistance	Apparent resistivity
C P P C or P C C P	$(\pm) R^\alpha$	$\rho_a^\alpha = 2\pi s R^\alpha$
C C P P or P P C C	$(\pm) R^\beta$	$\rho_a^\beta = 6\pi s R^\beta$
C P C P or P C P C	$(\pm) R^\gamma$	$\rho_a^\gamma = 3\pi s R^\gamma$

may be assumed that there is no significant lateral change in resistivity. If, however, one of the values of apparent resistivity is rising while the other two are falling (or vice versa), then it is likely that expansion of the spread has crossed a lateral resistivity boundary. Any depth soundings based on the apparent resistivity curve at spacings greater than that at which this divergent trend appears will be unreliable.

The Lee partition array

A modified Wenner array, the Lee partition array, also allows some estimate of the lateral variability of the resistivity of the surface material to be made. An extra potential electrode is inserted at the centre point of the array, making it a five-electrode array. Potential difference and thus resistivity can then be measured between the centre and left and centre and right potential electrodes. The two measured values should be the same, in which case there is no significant lateral variation in the surface material resistance. If variation is detected but is of an acceptable magnitude, then a simple average value may be used. Where substantial variation is apparent it may be better to adopt the Schlumberger array. The apparent resistivity for the Lee partition array is given by

$$\rho_a = K\Delta V/I ,$$

where $K = 4\pi a$.

The Schlumberger array

The Schlumberger array is designed to measure resistivity by measurement of the potential gradient and it uses a pair of closely spaced potential electrodes between a widely spaced pair of current electrodes (Fig. 5.8a). This array may be used either symmetrically (Fig. 5.8b) or non-symmetrically (Fig. 5.8c). For vertical electrical sounding the current electrode spacing is increased while keeping the potential electrode spacing constant. With large current electrode spacings it is necessary to increase the potential electrode spacing in order to preserve a measurable potential difference across them. Whenever the potential electrode spacing is increased, the current electrode spacing should be reduced to give an overlapping range of data. This will show up any lateral variations in near-surface resistivity. A well tried sequence of spacings is given in Table 5.4.

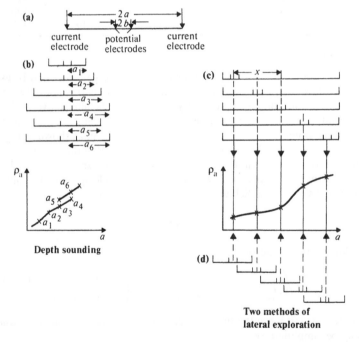

Figure 5.8 The Schlumberger array and its use for vertical sounding and two alternative methods of lateral exploration.

Table 5.4 Typical field layout for vertical electrical sounding using a Schlumberger array.

Observation number	Potential electrode spacing, 2b (m)	Current electrode spacing, 2a (m)	Observation number	Potential electrode spacing, 2b (m)	Current electrode spacing, 2a (m)
1	1.0	3.0	16	10	30
2	1.0	4.0	17	10	40
3	1.0	6.0	18	10	60
4	1.0	8.0	19	10	80
5	1.0	10.0	20	10	100
6	2.0	6.0	21	20	60
7	2.0	8.0	22	20	80
8	2.0	12.0	23	20	100
9	2.0	16.0	24	20	120
10	2.0	20.0	25	20	160
11	4	12	26	20	200
12	4	16	27	40	200
13	4	20	28	40	280
14	4	30	29	40	320
15	4	40	30	40	400

When used for horizontal profiling the Schlumberger array may be used in one of two ways. One method is to move the whole array laterally but to preserve a constant layout geometry (Fig. 5.8d). The alternative, which is generally quicker, is to use a wide spacing for the current electrodes and to move the potential electrodes laterally between them, as shown in Figure 5.8c. The latter method is subject to the limitation that the moving potential electrodes should not approach either of the current electrodes closer than about four or five times the potential electrode spacing. The apparent resistivity for the Schlumberger array is given by

$$\rho_a = K\Delta V/I .$$

For a non-symmetrical array,

$$K = 2\pi \left/ \left\{ \left(\frac{1}{a-x-b} - \frac{1}{a+x+b} \right) - \left(\frac{1}{a-x+b} - \frac{1}{a+x-b} \right) \right\} \right.$$

where a, x and b are the distances defined in Figure 5.8. This expression may be simplified to

$$K = \pi(a^2 - x^2)^2/2b(a^2 + x^2)$$

if the smallest distance between a current and potential electrode is always greater by a factor of ten or more than the distance between the two potential electrodes.

For a symmetrical Schlumberger array the apparent resistivity is given by

$$\rho_a = K\Delta V/I ,$$

where $K = \pi a^2/2b$.

Choice of an electrode configuration

Although, in general, both of the main electrode configurations will give satisfactory results under good ground conditions, there are various advantages to each method.

For vertical electrical sounding the Wenner array is mathematically simpler, measures larger values of the potential difference and thus gives a more accurate value for resistivity. It also has the advantage of a simple standard field procedure. Against this it does suffer from the disadvantage that all the electrodes must be moved between readings. Quite apart from the extra labour required, this means that the method is unable to distinguish between resistivity variations caused by deep layers and those caused by near-surface effects unless electrode switching is employed (Fig. 5.9). In terms of interpretation, complete curve matching or Tagg's method are the only ones that give acceptable results.

The Schlumberger array is theoretically less sound, in that to match theory with practice the potential electrode spacing should be infinitely small. However, the method reflects the electrical properties of the various layers well, provided that appropriately small potential electrode spacings are used. Table 5.4 provides guidance on this question. The reduction of working time yielded by the smaller number of electrodes moves is an important advantage of the Schlumberger array. Probably the most important advantage of the method is its ability to resolve resistivity variations caused by both near-surface and deep features. The effect of a change in surface material resistivity is to displace an arc of the sounding curve (Fig. 5.9) normally at a point where potential electrodes have been moved. This segment of the curve can be moved back into line with the rest of the curve to remove the near-surface effect. A major advantage of the Schlumberger array is that both curve matching and partial curve matching using auxiliary curves (see Keller & Frischknecht 1966) may be used to give a preliminary field interpretation of the data which is sufficiently accurate to plan further work.

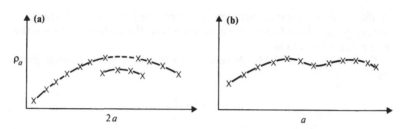

Figure 5.9 The effect of near-surface resistivity variations on the shape of (a) Schlumberger and (b) Wenner sounding curves.

For horizontal profiling the asymmetrical Schlumberger array or a half-Schlumberger array (see Telford *et al.*, 1976 for details) are the most efficient in terms of working time. They also reproduce the sense and location of anomalies well, especially if the potential electrode spacing is small with respect to the lateral extent of the anomaly. On the other hand, for laterally extensive anomalies the Wenner or half-Wenner array is preferable in that the observed resistivity curves are more symmetrical and the maximum observed resistivity contrast is never greater than the actual contrast.

Horizontal profiling (used on its own) is not particularly effective and should be calibrated wherever possible by vertical soundings and drilling.

5.9   Interpretation of resistivity sounding curves

Brief outlines of two methods of interpreting sounding curves for simple two-layer situations are given in the sections that follow. The curve-fitting method is relatively quick and simple but gives no indication of the accuracy of the results. Tagg's method (1934), although rather more time consuming, does not have this deficiency but does require that the resistivity of the top layer is accurately determined and that due account is taken of lateral and vertical variations within it.

Three-layer situations may produce a wide variety of curve types depending on the resistivity contrast between layers and the relative thicknesses of the layers. Complete curve matching is the preferred method of interpretation but it requires a bulky compendium of master curves which could not be accommodated in this volume. The principal compendia of curves are provided by Mooney and Wetzel (1956), Orellana and Mooney (1966), La Compagnie Générale de Géophysique (1963) and Rijkswaterstaat (1980).

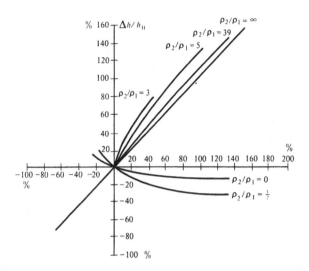

Figure 5.10 Percentage error in the interpreted thickness of a surface layer due to error in selecting the value of surface layer resistivity as a function of resistivity ratio (Keller & Frischknecht 1966).

Partial curve fitting can, however, provide a field interpretation that is adequate to enable the operator to plan further work. Partial curve fitting also has the advantage that it requires a minimum of extra master curves over and above those required for the simple two-layer situations. Full details of the method may be found in Keller and Frischknecht (1966).

Full curve matching may be used for data from both the Schlumberger and Wenner arrays using the appropriate set of master curves. Partial curve matching should be used only for interpreting three-layer Schlumberger soundings. Three-layer data from Wenner soundings should be interpreted using an atlas of curves using the whole-curve fitting approach.

One possible source of error in depth estimations is the incorrect selection of a value for the resistivity of the surface material. An assessment of the percentage error in depth determination as a function of the percentage error in surface resistivity may be made by reference to Figure 5.10.

Curve fitting for two-layer curves

Step 1: The field data are plotted on a transparent sheet using the same bilogarithmic scale as the appropriate set of master curves (Figs 5.11 or 12).

Step 2: Lay the transparent plot of the field data curve over the appropriate set of master curves and, ensuring that the axes remain parallel, move the field curve around until a satisfactory match with some portion of a master curve is obtained. Mark on the field curve overlay the origin of the matched master curves, the point at which $\rho_a = 1.0$ and $a = 1.0$ (Fig. 5.13a).

Step 3: Read off the values of the field data that correspond to the origin of the master curve, ρ_1. The value on the spacing axis is the thickness of the uppermost layer, h_0, and that on the apparent resistivity axis is the resistivity of the uppermost layer, ρ_0.

Step 4: Calculate the resistivity of the second layer by first converting the value of k to a resistivity ratio ρ_1/ρ_0 using Table 5.5 or the formula given in it. From this and the value of ρ_0 derived in Step 3, ρ_1 can be calculated.

Table 5.5 Values of ρ_1/ρ_0 corresponding to various values of k. For other values use $k = \rho_1 - \rho_0/\rho_1 + \rho_0$ or $\rho_1/\rho_0 = (k + 1)/(1 - k)$.

k	ρ_1/ρ_0		k	ρ_1/ρ_0	
	k positive	k negative		k positive	k negative
1.00	∞	0	0.45	2.64	0.38
0.95	39.00	0.03	0.40	2.33	0.43
0.90	19.00	0.05	0.35	2.08	0.48
0.85	12.33	0.08	0.30	1.86	0.54
0.80	9.00	0.11	0.25	1.67	0.59
0.75	7.00	0.14	0.20	1.50	0.67
0.70	5.67	0.18	0.15	1.35	0.74
0.65	4.71	0.21	0.10	1.22	0.81
0.60	4.00	0.25	0.05	1.11	0.90
0.55	3.44	0.29	0	1.00	1.00
0.50	3.00	0.33			

Tagg's method for two-layer curves

Step 1: The recorded value of apparent resistivity over very short electrode spacings may be taken as the resistivity of the uppermost layer, ρ_0.

Step 2: For some particular data point or value of apparent resistivity taken from a smoothed field curve, convert the value of apparent resistivity to one of apparent resistivity divided by surface resistivity, ρ_a/ρ_0. For this value of ρ_a/ρ_0 and its corresponding value for

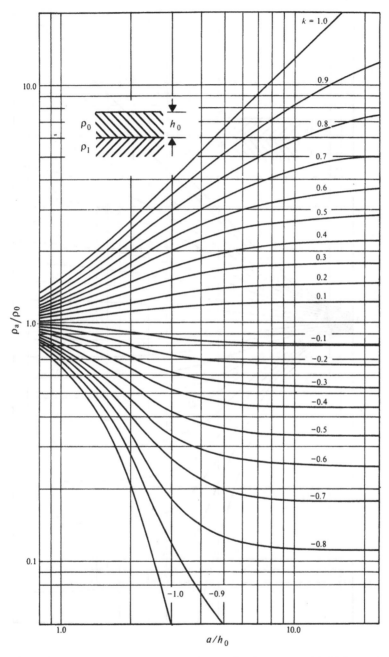

Figure 5.11 Two-layer master curves of normalised apparent resistivity against electrode spacing for the Wenner array.

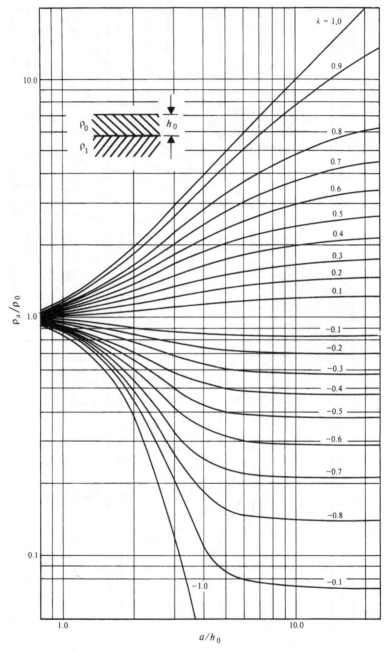

Figure 5.12 Two-layer master curves of normalised apparent resistivity against electrode spacing for the Schlumberger array.

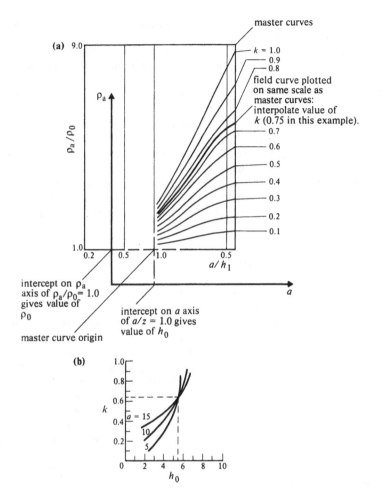

Figure 5.13 Method of use of master curves to determine the values of k, ρ_0 and h_0.

electrode spacing, a, read off from the master curves (Figs 5.11 or 12 depending on the electrode configuration used) all the possible values of h_0/a and k. Since the value of a is known, the value of h_0 can be calculated for each value of k. Plot a curve of h_0 against k for that data point (Fig. 5.13b).

Step 3: Repeat Step 2 for other data points to give a family of curves of h_0 against k (Fig. 5.13b).

Step 4: The intersection of these curves occurs at the required values of h_0 and k (Fig. 5.13b).

Step 5: Convert the value of k, the resistivity contrast factor, to a resistivity ratio ρ_1/ρ_0 using Table 5.5 or the formula given in it. Calculate the resistivity of the second layer using the value of ρ_0 determined in Step 1 and the resistivity ratio ρ_1/ρ_0.

In most cases the curves will not intersect perfectly at a single point but will describe an area encompassing the range of possible values of k and h_0. Step 1 is particularly crucial and great care should be taken to ensure that a realistic value of ρ_0 is obtained. In particular, care must be taken not to use electrode separations of such length as to allow current penetration into the lower medium, while ensuring that excessively short separations do not give a false value based only on the topmost layers of material.

5.10 Resistivity profiling

Horizontal resistivity profiling or electrical mapping is a useful method of conducting overburden surveys. It is, however, strongly dependent upon the electrode configuration chosen in terms of its resolution. Reference should be made to Section 5.8 for some guidance on this topic. Some prior knowledge of the geological situation is desirable before horizontal profiling is carried out and this may be obtained by using a number of vertical soundings. It is then possible to map the depth or thickness of a specific layer using the technique set out in Section 5.11. In general terms, the horizontal profiling method is best reserved for this type of activity and it should be restricted to situations where the resistivity contrasts are high and the anomalous bodies are laterally extensive. The method does not 'see' three-dimensional bodies of small extent very efficiently. Figures 5.14 and 15 give some examples of resistivity profiles across various simple geological structures.

5.11 Overburden surveys

A common use of resistivity surveys is to determine the thickness of a conducting overburden over an insulating bedrock. This can be achieved efficiently by running a horizontal profile using two fixed electrode separations. The chosen shorter separation should measure the resistivity of the overburden, however thin it may be. The longer separation is chosen

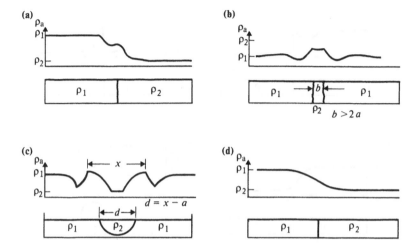

Figure 5.14 Some examples of resistivity profiles over geological structures using the full Wenner electrode configuration. (a) Vertical contact with the electrode spread aligned normal to the contact. (b) Vertical dyke of width greater than twice the electrode separation. (c) Hemispherical sink of width greater than twice the electrode separation; diameter of the sink can be determined approximately as the distance between the 'horns' of the field curve minus the electrode spacing (after Van Nostrand & Cook 1954). (d) Vertical contact with the electrode spread aligned parallel to the contact and being moved normal to it.

to measure the bedrock resistivity, even in areas where the overburden is thickest.

The thickness of the overburden (T_s) is given by

$$T_s = \rho_1 S_1 \,,$$

where ρ_1 = overburden resistivity measured with the shorter electrode spacing, S_1 = overburden conductivity = a_2/ρ_2, a_2 = the longer electrode spacing, ρ_2 = bedrock resistivity measured at the longer electrode spacing.

This method should be calibrated wherever possible by vertical soundings.

5.12   Electrical properties of Earth materials

Electrical conduction in the majority of rocks and sediments is electrolytic

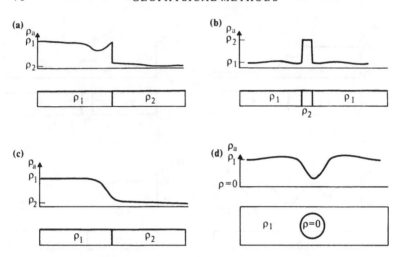

Figure 5.15 Some examples of resistivity profiles over geological structures using the full Schlumberger electrode configuration. (a) Vertical contact with the electrode spread aligned normal to the contact. (b) Vertical dyke of width of the order of half the total length of the electrode spread. (c) Vertical contact with the electrode spread aligned parallel to the contact. (d) Perfectly conducting buried sphere. This type of anomaly may be very weakly developed unless the buried sphere is close to the surface.

because, with the exception of a relatively few metallic minerals, most minerals are insulators. Conduction therefore takes place essentially through the interstitial waters and is to a large extent a function of the degree of porosity, water content and the quantity of dissolved salts in the water. The most important exceptions to this general rule are the metallic sulphides and materials rich in clay minerals, both of which are conductors. Crystalline rocks with low porosities conduct electricity through water-filled cracks and fissures, and the degree of fissuring is the main control on the resistivity of such rocks unless the waters are saline. In porous rocks and sediments the degree of saturation is often the controlling factor on the value of resistivity (Table 5.6).

Resistivity is thus a highly variable property and it varies not only between formations but also within the formation. This is especially true of near-surface unconsolidated materials such as drift deposits. Table 5.7, which gives a set of typical values of resistivity, should be used with these general points in mind.

Table 5.6 Variation of resistivity with water content (Telford *et al.* 1976).

Rock	$\%H_2O$	$\rho\ (\Omega\ m)$	Rock	$\%H_2O$	$\rho\ (\Omega\ m)$
siltstone	0.54	1.5×10^4	pyrophyllite	0.76	6×10^6
siltstone	0.44	8.4×10^6	pyrophyllite	0.72	5×10^7
siltstone	0.38	5.6×10^8	pyrophyllite	0.7	2×10^8
coarse grain SS	0.39	9.6×10^5	pyrophyllite	0	10^{11}
coarse grain SS	0.18	10^8	granite	0.31	4.4×10^3
medium grain SS	1.0	4.2×10^3	granite	0.19	1.8×10^6
medium grain SS	1.67	3.2×10^6	granite	0.06	1.3×10^8
medium grain SS	0.1	1.4×10^8	granite	0	10^{10}
graywacke SS	1.16	4.7×10^3	diorite	0.02	5.8×10^5
graywacke SS	0.45	5.8×10^4	diorite	0	6×10^6
arkosic SS	1.26	10^3	basalt	0.95	4×10^4
arkosic SS	1.0	1.4×10^3	basalt	0.49	9×10^5
organic limestone	11	0.6×10^3	basalt	0.26	3×10^7
dolomite	2	5.3×10^3	basalt	0	1.3×10^8
dolomite	1.3	6×10^3	olivine–pyroxene	0.028	2×10^4
dolomite	0.96	8×10^3	olivine–pyroxene	0.014	4×10^5
peridotite	0.1	3×10^3	olivine–pyroxene	0	5.6×10^7
peridotite	0.03	2×10^4			
peridotite	0.016	10^6			
peridotite	0	1.8×10^7			

SS = sandstone.

Table 5.7 Resistivities of rocks and sediments (Telford *et al.* 1976).

Rock type	Resistivity range $(\Omega\ m)$
granite	3×10^2–10^6
granite porphyry	4.5×10^3 (wet)–1.3×10^6 (dry)
feldspar porphyry	4×10^3 (wet)
albite	3×10^2 (wet)–3.3×10^3 (dry)
syenite	10^2–10^6
diorite	10^4–10^5
diorite porphyry	1.9×10^3 (wet)–2.8×10^4 (dry)
porphyrite	1.5×10^4 (wet)–3.3×10^3 (dry)
carbonatised porphyry	2.5×10^3 (wet)–6×10^4 (dry)
quartz porphyry	3×10^2–9×10^5
quartz diorite	2×10^4–2×10^6 (wet)–1.8×10^5 (dry)
porphyry (various)	60–10^4
dacite	2×10^4 (wet)
andesite	4.5×10^4 (wet)–1.7×10^2 (dry)
diabase porphyry	10^3 (wet)–1.7×10^5 (dry)
diabase (various)	20–5×10^7
lavas	10^2–5×10^4
gabbro	10^3–10^6
basalt	10–1.3×10^7 (dry)
olivine norite	10^3–6×10^4 (wet)
peridotite	3×10^3 (wet)–6.5×10^3 (dry)

Table 5.7 – *continued.*

Rock type	Resistivity range (Ω m)
hornfels	8×10^3 (wet)–6×10^7 (dry)
schists (calcareous and mica)	20–10^4
tuffs	2×10^3 (wet)–10^5 (dry)
graphite schist	10–10^2
slates (various)	6×10^2–4×10^7
gneiss (various)	68×10^4 (wet)–3×10^6 (dry)
marble	10^2–2.5×10^8 (dry)
skarn	2.5×10^2 (wet)–2.5×10^6 (dry)
quartzites (various)	10–2×10^8
consolidated shales	20–2×10^3
argillites	10–8×10^2
conglomerates	2×10^3–10^4
sandstones	1–6.4×10^8
limestones	50–10^7
dolomite	3.5×10^2–5×10^3
unconsolidated wet clay	20
marls	3–70
clays	1–100
alluvium and sands	10–800
oil sands	4–800

6 The description of landscape-forming materials

The purpose of this chapter is to provide a framework and the necessary aids for the description of the main landscape-forming materials: rocks, unconsolidated sediments and soils. Space does not permit any useful material on mineral identification to be included and the user is therefore recommended to provide some appropriate mineralogy text such as Rutley (1970) as a complement to this section. Section 6.1 provides for the naming of rock types in the field and follows the scheme used by the Soil Survey of England and Wales. As far as the geomorphologist is concerned, probably one of the most important aspects of the geology of an area is the durability or strength of individual rock types. Two schemes for the description of rock-mass strength and character are given in Section 6.2. No attempt has been made to integrate the two schemes since this would impair cross comparison between results already published using each scheme. One of the requirements of rock-mass classification is a reproducible measure of intact rock strength. Few strength tests use equipment that is portable and used in the field. Two exceptions to this rule are briefly explained in Section 6.3. Section 6.4 deals with the description of soils (in the pedological rather than the engineering sense) and the layout adopted follows closely that used in the *Soil survey field handbook* (Hodgson 1974). Section 6.5 is concerned with the description of sediments and sedimentary rocks as much from the point of view of the sedimentologist as from that of the geomorphologist. Throughout this chapter, key features or characteristics are in **bold** type to facilitate scanning as a check of features to be noted or described.

6.1 Rock identification

It is rarely possible to name a rock precisely in the field without recourse to thin-section examination. Close examination with a hand lens should allow the specimen to be assigned to one of the broad categories given in

Tables 6.1, 2, 3 and 4 for igneous, sedimentary, metamorphic and pyro-clastic rocks respectively. In all cases the abundance of any significant accessory minerals should be estimated and the appropriate prefix given to the broad rock name. Charts for the estimation of abundance are given in Chapter 13. In the case of pyroclastic rocks the rock name should be prefixed by a compositional term, e.g. a *rhyolitic* tuff. An alternative petrographic nomenclature is given as part of the scheme for the description of rock material in Table 6.5. A more detailed scheme for the description of sediments and sedimentary rocks is given in Section 6.5.

Table 6.1 Nomenclature of igneous rocks. The relative abundance of each essential mineral in each rock type is indicated by the vertical extent of the field for that mineral in the diagram.

	Acid	*Intermediate*		*Basic*	*Ultra-basic*
grain size (mm)					
coarse (>5)	granite	syenite	diorite	gabbro	peridotite
medium (1–5)	microgranite	microsyenite	microdiorite	dolerite	
fine (<1)	rhyolite	trachyte	andesite	basalt	

essential minerals

density (g cm^{-3})	2.4–2.7	2.8	2.8	2.9	>3.0
accessory minerals	micas	biotite	biotite	augite	
	hornblende	augite	augite	olivine	
		hornblende	hornblende		
		Na plagioclase			
colour	pale or white	intermediate		dark	very dark or black

6.2 Rock mass description

Two schemes for the description of rock masses are included in this section. The first, that of the Geological Society Engineering Group Working Party (1977), emphasises the detailed description of discontinuities while

Table 6.2 Nomenclature of sedimentary rocks (Hodgson 1974).

Grain size	Hardness	Mineralogy and fabric	Rock type	Rock subtype based on accessory minerals or impurities
amorphous–microcrystalline	very hard	chalcedonic silica often with conchoidal fracture	flint	
		chalcedony with occasional fine radio-laria or sponge spicules	chert	
		quartz grains with continuous silica cement	quartzite	
clayey (<2 μm)	hard–soft	predominantly clay-size particles consolidated, without fissility	mudstone	calcareous
		weakly cleaved varieties transitional to slates	cleaved mudstone	micaceous haematitic (red) carbonaceous (black)
		predominantly clay-size particles consolidated with fissility	clay shale	ferruginous
silty (2–60 μm)	hard–soft	predominantly silt-size particles consolidated, without fissility	siltstone	
		predominantly silt-size particles consolidated with fissility	silty shale	
sandy (60 μm–2 mm)	very hard	silica-cemented sandstone transitional to quartzite	quartzitic sandstone	calcareous
	hard–soft	predominantly sand-size particles with siliceous, calcareous, ferruginous or other cements	sandstone	micaceous ferruginous feldspathic glauconitic dolomitic
		predominantly sand-size particles with siliceous, calcareous, ferruginous or other cements with a proportion of particles >2 mm	grit	

Table 6.2 – continued.

Grain size	Hardness	Mineralogy and fabric	Rock type	Rock subtype based on accessory minerals or impurities
mixed	hard	mixed sand, silt and clay-size particles of varying mineralogy including quartz, feldspars, clay minerals and rock fragments	greywacke	
coarse (>2 mm)	hard–soft	angular rock fragments cemented in a finer matrix	breccia	subtypes based on nature of rock fragments and cement
		rounded stones cemented in a finer matrix	conglomerate	
carbonatic (>50% CaCO$_3$)	hard–soft	predominantly calcium carbonate or undifferentiated components	limestone	micaceous ferruginous glauconitic dolomitic sandy argillaceous
		predominantly calcium carbonate with abundant shell remains	shelly limestone	
		predominantly calcium carbonate with abundant ooliths	oolitic limestone	
		predominantly calcium carbonate with abundant pisoliths	pisolitic limestone	
	hard	predominantly calcium carbonate of microcrystalline (or muddy) form	calcite mudstone	
	soft	predominantly calcium carbonate of microcrystalline (or muddy) form	chalk	
carbonaceous	soft–hard	black carbonised carbonaceous material	coal	

Table 6.3 Nomenclature of metamorphic rocks (Hodgson 1974).

	Normal fabric	*Main minerals*	*Rock name*
unfoliated	hornfelsic (fine-graded mosaic of unorientated grains)	micas, garnet, pyroxenes, etc.	hornfels
	granulose (medium-grained, equidimensional mineral grains)	garnet, epidote, diopside, etc.	skarn
		calcite or dolomite	marble
		quartz	quartzite
foliated (flaky minerals occurring in layers)	poorly foliated	hornblende, feldspar	amphibolite
		quartz or pyroxene, feldspar	granulite
		hypersthene, quartz, feldspar	charnockite
		quartz, feldspar, chlorite	argillite
	well developed cleavage planes	quartz, micas, chlorite	slate
	cleavage planes becoming schistose	quartz, micas, chlorite	phyllite
	schistose (finely foliated)	micas, chlorite, quartz	mica schist
		calcite, micas, quartz	calc schist
		chlorite, actinolite, epidote	chlorite schist
		talc	talc schist
		glaucophane	glaucophane schist
		hornblende, feldspar	hornblende schist
		garnet, micas, chlorite	garnet schist
		staurolite, micas	staurolite schist
		sillimanite	sillimanite schist
		graphite, micas, quartz	graphite schist
	gneissic (coarsely foliated or banded)	quartz, feldspar, micas	gneiss
	injected with granite	quartz, feldspar	migmatite
brecciated	mylonitic (fine-grained and laminated)	quartz, feldspar	mylonite
	augen (almond-shaped eyes of quartz and feldspar)	'eyes' of quartz and feldspar	augen gneiss

Table 6.4 Nomenclature of pyroclastic and non-crystalline igneous rocks (Hodgson 1974).

Fragment size	Name
very coarse (>32 mm)	agglomerate
coarse (4–32 mm)	volcanic breccia
fine (<4 mm)	tuff
non-crystalline	glass (vitreous)
	pumice (frothy)

that of Selby (1980) is a rock-mass strength rating scheme designed specifically for use by geomorphologists. No attempt has been made to integrate the two schemes, and the user must choose the one that is most appropriate to his own needs. Where necessary, additional information can be included by using the appropriate parts of the descriptive schemes offered for soils and sediments in Sections 6.4 and 5 respectively. Both schemes advocate the use of field and/or laboratory testing to determine parameters such as the strength of intact rock, permeability, etc. Most of these tests are outside the scope of this handbook in that they require procedures, charts and formulae specific to the equipment being used. The point load test and the Schmidt hammer may be used for field determination of intact rock strength and brief instructions and charts for the latter are provided in Section 6.3. Also in Section 6.3 are some notes on the use of the point load test.

The Geological Society Engineering Group Working Party scheme

This scheme is based on the identification, description and quantification of 'index' properties of both the intact rock material and the overall rock mass. Each category of each of the index properties is numbered and the number may be used to code the data on a data recording sheet. Some examples of these may be found in the original paper or they may be designed specifically for the user's own purposes. Qualifying statements should be added wherever the terms given prove inadequate. The index properties for intact rock material are rock type, rock colour, grain size, strength, texture and fabric.

Rock type should be identified using the petrographic names given in Table 6.5. A simple scheme for the description of **rock colour** is given in

Table 6.6. A colour chosen from Column 3 of the table may be complemented by appropriate terms from Columns 1 and 2. The **grain size** terms in common use for both rocks and soils should be used and these are given in Table 6.7. **Intact rock material strength** should ideally be determined by using the unconfined compressive strength test, the point load test or the Schmidt hammer. In the absence of such test data an estimate of the rock strength can be arrived at by using Table 6.8. Section 6.3 gives guidance on the Schmidt hammer and the point load tests. Qualifying notes on the **texture and fabric** of the rock material should be made and particular attention paid to any preferred orientation of the constituent grains.

The index properties of the rock mass used in this scheme are the nature and three-dimensional spacing of discontinuities, weathering and alteration. A discontinuity is considered to be a plane in the rock across which the rock has low or zero tensile strength, although it need not be a plane of separation. Where possible it is desirable to differentiate between **discontinuity types** (Table 6.9), especially between those that are unique, e.g. faults, and those that occur in sets, e.g. joints, cleavages and bedding planes. Sufficient information should be recorded, preferably on a map or plan, to **locate each discontinuity in space** and the **dip amount and direction** should be recorded for each discontinuity. The **spacing** between adjacent discontinuities should be measured by counting the number that cut a traverse of known length and the results expressed as a mean and a range. Ideally the measurements should be taken along three mutually perpendicular axes and categorised using the terms in Table 6.10.

The **aperture** or separation across discontinuities should be measured and described using terms given in Table 6.11. Any **material infilling discontinuities** should be identified and described using either the terms for rock material in Table 6.5 or appropriate terms from Section 6.4 for soils. The nature of infilling material may be classified using the terms in Table 6.12. The **thickness of infilling material** is taken as that thickness separating the host rock surfaces and may be categorised using the terms in Table 6.11. Some indication of the **persistence of discontinuity surfaces** is required and it is recommended that the maximum trace length be measured. Notes should be made as to whether the total trace length can be seen and whether the discontinuity terminates against solid rock or against another discontinuity.

The nature of the discontinuity surfaces should be recorded. **Waviness** refers to the first-order asperities that appear as undulations on the surface and which would be unlikely to shear off during any movement. They can be described by using a tape or rule to measure mean amplitude and

Table 6.5 Rock type classification (Geological Society Engineering Group Working

Genetic group	Detrital sedimentary		Pyroclastic	Chemical/ Organic
usual structure	bedded		bedded	
composition	grains of rock, quartz, feldspar and minerals	at least 50% of grains are of carbonate	at least 50% of grains are of fine-grained volcanic rock	
Grain size scale (mm)				
very coarse grained 60	grains are of rock fragments		rounded grains: **agglomerate** (31)	
	rounded grains: **conglomerate** (10)	**calcirudite** (21)		
coarse grained 2	angular grains: **breccia** (11)		angular grains: **volcanic breccia** (32)	
				saline rocks: halite (41) anhydrite (42) gypsum (4.3)
medium grained	**sandstone:** grains are mainly mineral fragments			
	quartz sandstone: 95% quartz, voids empty or cemented (12)			
	arkose: 75% quartz, up to 25% feldspar; voids empty or cemented (13)	**calc-arenite** (22) **tuff** (33)		
	argillaceous sandstone: 75% quartz, 15% + fine detrital material (14)		**fine-grained tuff** (34) chert (44)	
fine grained 0.06	**mudstone** (15) **shale:** fissile mudstone (16)	**calcisiltite** (23)		**flint** (45)
	siltstone: 50% fine-grained particles (17)	**calcilutite** (24)	very fine-grained tuff (35)	**coal** (46) others (47) specify
very fine grained 0.002	**claystone:** 50% very fine-grained particles (18) **calcareous mudstone** (19)			
glassy				

(rudaceous / arenaceous / argillaceous or lutaceous — grain size scale column; limestone (undifferentiated))

Metamorphic		Igneous			
foliated quartz, feldspars, micas, acicular dark minerals	massive	massive light coloured minerals are quartz feldspar, mica and feldspar-like minerals			dark minerals
		acid rocks	intermediate rocks	basic rocks	ultrabasic rocks
Migmatite (51)	**hornfels** (61)	**pegmatite** (81)			**pyroxenite** (01) **peridotite** (02)
		granite (71)	**diorite** (82)	**gabbro** (92)	
gneiss: alternate layers of granular and flakey minerals (52)	**marble** (62)	—			**serpentine** (03)
schist (53)	**granulite** (63) **quartzite** (64)	**microgranite** (72)	**microdiorite** (83)	**dolerite** (93)	
phyllite (54)	**amphibolite** (65)				
slate (55)					
		rhyolite (73)	**andesite** (84)	**basalt** (94)	
mylonite (56)					
		obsidian (74)	**pitchstone** (85)	**tachylyte** (95)	

Table 6.6 Descriptive terms for rock colour (Geological Society Engineering Group Working Party 1977).

Column 1			Column 2			Column 3	
Term	Code		Term	Code		Term	Code
light	1		pinkish	1		pink	1
dark	2		reddish	2		red	2
			yellowish	3		yellow	3
			brownish	4		brown	4
			olive	5		olive	5
			greenish	6		green	6
			bluish	7		blue	7
						white	8
			greyish	8		grey	9
						black	0

Table 6.7 Descriptive terms for grain size (Geological Society Engineering Group Working Party 1977).

Term	Particle size	Equivalent soil grade	Category	
very coarse-grained	>60 mm	boulders and cobbles	1	visible to naked eye
coarse-grained	2–60 mm	gravel	2	visible to naked eye
medium-grained	60 μm–2 mm	sand	3	
fine-grained	2–60 μm	silt	4	not visible
very fine-grained	<2 μm	clay	5	to the naked eye

Table 6.8 Descriptive terms for rock strength (Geological Society Engineering Group Working Party 1977).

Term	Unconfined compressive strength $(MN\ m^{-2} = MPa)$	Field estimation of hardness	Category
very strong	>100	very hard rock – more than one blow of geological hammer required to break specimen	1
strong	50–100	hard rock – hand-held specimen can be broken with a single blow of geological hammer	2
moderately strong	12.5–50	soft rock – 5 mm indentations with sharp end of pick	3
moderately weak	5.0–12.5	too hard to cut by hand into a triaxial specimen	4
weak	1.25–5.0	very soft rock – material crumbles under firm blows with the sharp end of a geological pick	5
very weak rock or hard soil	0.60–1.25	brittle or tough, may be broken in the hand with difficulty	6
very stiff	0.30–0.60	soil can be indented by the fingernail	7
stiff	0.15–0.30	soil cannot be moulded in fingers	8
firm	0.08–0.15	soil can be moulded only by strong pressure of fingers	9
soft	0.04–0.08	soil easily moulded with fingers	0

Table 6.9 Types of discontinuity (Geological Society Engineering Group Working Party 1977).

Category	Type	Category	Type
0	fault zone	5	shear
1	fault	6	fissure
2	joint	7	tension crack
3	cleavage	8	foliation
4	schistosity	9	bedding

Table 6.10 Terms for discontinuity spacing (Geological Society Engineering Group Working Party 1977).

Term	Spacing	Category
extremely wide	>2 m	1
very wide	600 mm–2 m	2
wide	200–600 mm	3
moderately wide	60–200 mm	4
moderately narrow	20–60 mm	5
narrow	6–20 mm	6
very narrow	<6 mm	7

Table 6.11 Descriptive terms for discontinuity aperture and thickness (Geological Society Engineering Group Working Party 1977).

Term	Aperture or thickness	Category
wide	>200 mm	1
moderately wide	60–200 mm	2
moderately narrow	20–60 mm	3
narrow	6–20 mm	4
very narrow	2–6 mm	5
extremely narrow	0–2 mm	6
tight	zero	7

Table 6.12 Descriptive terms for the nature of discontinuity infill (Geological Society Engineering Group Working Party 1977).

Nature of infill	Category	Nature of infill	Category
clean	1	swelling clay or clay matrix	5
surface staining only	2	cemented	6
non-cohesive	3	chlorite, talc or gypsum	7
inactive clay or clay matrix	4	other – specify	8

Table 6.13 Descriptive terms for discontinuity roughness (Geological Society Engineering Group Working Party 1977).

Degree of roughness	Category	Degree of roughness	Category
polished	1	defined ridges	5
slickensided	2	small steps	6
smooth	3	very rough	7
rough	4		

Table 6.14 Descriptive terms for block shape (Geological Society Engineering Group Working Party 1977).

Term	Description	Category
blocky	approximately equidimensional	1
tabular	one dimension considerably less than the other two	2
columnar	one dimension considerably larger than the other two	3

Table 6.15 Descriptive terms for block size (Geological Society Engineering Group Working Party 1977).

Term	Block size	Equivalent discontinuity spacing in blocky rock	Category
very large	>8 m^3	extremely wide	1
large	0.2–8 m^3	very wide	2
medium	0.008–0.2 m^3	wide	3
small	0.0002–0.008 m^3	moderately wide	4
very small	<0.0002 m^3	less than moderately wide	5

Table 6.16 Descriptive terms for weathering grades (Geological Society Engineering Group Working Party 1977).

Term	Description	Grade	Category
fresh	no visible sign of rock material weathering	1A	1
faintly weathered	discoloration on major discontinuity surfaces	1B	1
slightly weathered	discoloration indicates weathering of rock material and discontinuity surfaces; all the rock material may be discoloured by weathering and may be somewhat weaker than in its fresh condition	II	2
moderately weathered	less than half of the rock material is decomposed and/or disintegrated to a soil; fresh or discoloured rock is present either as a continuous framework or as corestones	III	3
highly weathered	more than half the rock material is decomposed and/or disintegrated to a soil; fresh or discoloured rock is present either as a discontinuous framework or as corestones	IV	4
completely weathered	all rock material is decomposed and/or disintegrated to soil; the original mass structure is still largely intact	V	5
residual soil	all rock material is converted to soil; the mass structure and material fabric are destroyed; there is a large change in volume, but the soil has not been significantly transported	VI	6

wavelength. **Surface roughness** can be described using the terms in Table 6.13, although these terms are meaningful only when the direction of trend of the irregularities is least resistant to sliding. In order to maintain uniformity of assessment it is recommended that typical examples of each category at each site be photographed.

Any other information about discontinuity nature that may be relevant should also be included. Such information may include the nature of **water seepage**, the strength of weathered rock adjacent to discontinuities, the presence of swelling materials and, most importantly, any evidence of recent movement or **instability**.

Discontinuity spacing in three dimensions may be described by using the terms given in Table 6.14 for the shape of blocks formed by the intersection of discontinuities. In addition the orientation of the long and short dimensions should be specified. Block size may be described using the terms in Table 6.15.

The **weathering** of rock mass may be described in terms of the distribution of the weathered materials within it and the effect of weathering on discontinuities. Descriptive terms for weathering grade are given in Table 6.16 but they may need to be modified to suit particular situations. The **distribution of weathering** grades within the rock mass should be noted.

Rock mass rating (Selby 1980)

This scheme is intended for the assessment of the mass strength of rock in the field using only limited equipment. The items of equipment needed are a tape measure, compass-clinometer and a strength tester of some type, although the last item may be dispensed with. Each parameter of the classification is given a rating value (r) and the rock-mass strength is indicated by the sum of the ratings. Seven parameters are used in the scheme and the ratings for each category of each parameter are obtained from tables.

The outcrop under examination should be divided into internally similar areas and the overall rating determined for each of these areas. Measurements of discontinuity spacing and continuity should be made on both horizontal and vertical faces and the area of face required for this is usually of the order of 10 m^2. As well as assigning the rating values to each unit in turn it is recommended that, at least initially, detailed descriptions and field sketches be made in order to check on the consistency of application. The **intact rock strength** should ideally be determined by measurement of the unconfined uniaxial compressive strength. Other suitable methods are the use of the point load test or the Schmidt hammer, or by reference to the simple tests outlined in Table 6.17. Section 6.3 gives guidance on the use of the Schmidt hammer and point load tests. **Weathering grade** may be determined by reference to Table 6.18. Ratings of less than three for the residual soil are rarely needed since such materials are usually covered by vegetation.

Ratings are given for various **discontinuity configurations** in Table 6.19, but they should be treated only as an approximate guide and the user should be guided in their use by local field evidence, particularly any evidence of previous slope failure. Table 6.20 gives ratings for different discontinuity spacings. Superficial fissuring should be ignored and only the major deeply penetrating discontinuities considered. Reference should also be made to Table 6.20 for a rating value for **discontinuity width**. The width to be recorded is that of the discontinuity at a depth of at least

Table 6.17 Description, classification and rating of the strength of intact rock (Selby 1980).

Description	Unconfined compressive strength (MPa)	Point load strength (MPa)	Approximate N-type Schmidt hammer 'R' value	Rating, 'r'
very weak rock, crumbles under sharp blows with geological pick, can be cut with a knife, e.g. chalk, rocksalt, lignite	1–25	0.04–1.0	10–35	5
weak rock, shallow cuts or scratches made with a sharp knife, pick point indents with firm blow, e.g. coal, siltstone, schist	25–50	1.0–2.0	35–40	10
moderately strong rock, knife cannot scrape surface, shallow indentation under firm blow from pick, e.g. slate, shale, sandstone, mudstone, ignimbrite	50–100	2.0–4.0	40–50	14
strong rock, hand sample breaks with one firm blow from the hammer end of geological pick	100–200	4.0–8.0	50–60	18
very strong rock, requires many blows from geological pick to break intact sample	>200	>8.0	>60	20

100 mm into the outcrop. Some judgement has to be exercised in assessing width; 'hairline' fractures should, for instance, be ignored. A conservative estimate is advocated in interpreting the classes in Table 6.20. Rating values for varying degrees of **fracture continuity** are also given in Table 6.20. Also included in this rating is an assessment of the effect of **infill**, since swelling pressures may reduce the strength across the discontinuity. An estimate should also be made of the **volume of water flowing**

Table 6.18 Description and rating of mass weathering grades (Selby 1980).

Grade	Class	Rating, 'r'	Description
VI	residual soil	<3	a pedological soil containing characteristic horizons and no sign of original rock fabric
V	completely weathered	3	rock is discoloured and changed to a soil but some original rock fabric and texture is largely preserved; some corestones or corestone ghosts may be present
IV	highly weathered	5	rock is discoloured throughout; discontinuities may be open and have discoloured surfaces and the fabric of the rock near to the discontinuities may be altered so that up to one half of the rock mass is decomposed and disintegrated to a stage in which it can be excavated with a geological hammer; corestones may be present but not generally interlocked
III	moderately weathered	7	rock is discoloured throughout most of its mass, but less than half of the rock mass is decomposed and disintegrated; alteration has penetrated along discontinuities which may be zones of weakly cemented alteration products or soil; corestones are fitting
II	slightly weathered	9	rock may be slightly discoloured, particularly adjacent to discontinuities which may be open and will have slightly discoloured surfaces; intact rock is not noticeably weaker than fresh rock
I	unweathered fresh rock	10	parent rock showing no discoloration, loss of strength or any other weathering effects

Table 6.19 Description, rating and classification for discontinuity orientations (Selby 1980).

	Rating, 'r'	Mode of discontinuity formation	
		Tensile (rough)	Shear (smooth)
very unfavourable	5	surfaces dip out of the slope: planar surfaces 30–80°; random surfaces >70°	surfaces dip out of the slope: planar surfaces 20°; random surfaces 30°
unfavourable	9	surfaces dip out of the slope: planar surfaces 10–30°; random surfaces 10–70°	surfaces dip out of the slope: planar surfaces 10–20°; random surfaces 10–30°
fair	14	horizontal to 10°dip out of the slope; nearly vertical (80–90°) in hard rocks with planar surfaces	horizontal to 10° dip out of the slope
favourable	18	surfaces dip from horizontal to 30° into the slope; cross joints not always interlocked	
very favourable	20	surfaces dip at more than 30° into the slope; cross joints are weakly developed and interlocking	

out of each 10 m² of the rock surface during the wettest season of the year. If this is not possible, a subjective term from Table 6.20 should be used.

6.3 Field methods for determining rock material strength

The Schmidt rebound hammer test

Both the 'N' and 'L' types of hammer have been used by geomorphologists and engineers to assess rock material strength. Day and Goudie (1977) have tested the type 'N' hammer, and Deere and Miller (1966) used the type 'L' instrument. The following precautions should be taken when using either instrument:

(a) Readings should be taken at least 60 mm away from edges or cracks.
(b) The hammer must be moved to a fresh spot for each reading.
(c) The test surface should be flat and free from flakes or dirt. However, when the instrument is being used to determine the strength of the

Table 6.20 Geomorphic rock mass strength classification and ratings (Selby 1980).

	1 Very strong	2 Strong	3 Moderate	4 Weak	5 Very weak
intact rock strength (N-type Schmidt hammer 'R')	100–60 very strong r = 20	60–50 strong r = 18	50–40 moderate r = 14	40–35 weak r = 10	35–10 very weak r = 5
weathering	unweathered r = 10	slightly weathered r = 9	moderately weathered r = 7	highly weathered r = 5	completely weathered r = 3
spacing of discontinuities	>3 m solid r = 30	3–1 m massive r = 28	1–0.3 m blocky/seamy r = 21	300–50 mm fractured r = 15	<50 mm crushed or shattered r = 8
joint orientations	very favourable; steep dips into slope, cross joints interlock r = 20	favourable; moderate dips into slope r = 18	fair; horizontal dips, or nearly vertical (hard rocks only) r = 14	unfavourable; moderate dips out of slope r = 9	very unfavourable; steep dips out of slope r = 5
width of joints	<0.1 mm r = 7	0.1 mm r = 6	1–5 mm r = 5	5–20 mm r = 4	>20 mm r = 2
fracture continuity	none continuous r = 7	few continuous r = 6	continuous, no infill r = 5	continuous, thin infill r = 4	continuous, thick infill r = 1
outflow of ground-water	none r = 6	trace r = 5	slight (<25 l min^{-1} from 10 m^2) r = 4	moderate (25–125 l min^{-1} from 10 m^2) r = 3	great (>125 l min^{-1} from 10 m^2) r = 1
total rating	100–91	90–71	70–51	50–26	<26

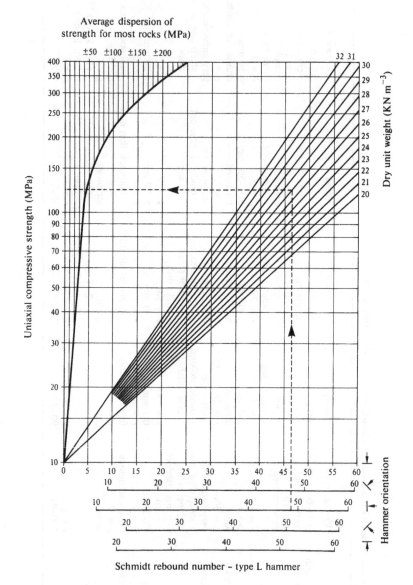

Figure 6.1 Relationship between hammer orientation, rebound numbers and uniaxial compressive strength for the type 'L' hammer (after Deere & Miller 1966).

wall material of discontinuities, no surface preparation should be carried out.

(d) The number of impacts made at any one site should be at least ten or fifteen. Selby (1980) reports that the most reliable results are obtained if the upper and lower twenty per cent of impact readings are ignored and measurements continued until the deviation from the mean value of the remainder does not exceed plus or minus three points.

(e) Corrections should be made for the attitude in which the hammer is used. Figure 6.1, based on that of Deere and Miller (1966), may be used both to correct for hammer impact angle and to convert the impact reading to uniaxial compressive strength for the 'L' type hammer. For the 'N' type hammer Table 6.21 gives the correction factors to be applied to the R value.

(f) The instrument should be regularly recalibrated against the test anvil supplied by the manufacturer.

(g) Readings will vary with the moisture content of the sample. Ideally, therefore, all readings should be made under the same moisture content conditions.

The conversion of rebound readings to unconfined uniaxial compressive strength has been criticised by Neville (1973, p. 501). For any major project using the Schmidt hammer it may be advisable to make one's own calibration curve by field test for R values and then test specimens for their unconfined uniaxial compressive strength by the standard engineering procedures.

Singer and Yaalon (1974) found the following linear regression between R and the compressive strength:

$$\log(\delta/9.81 \times 10^4) = 0.0387\,R + 0.826$$

Table 6.21 Correction of type 'N' test hammer rebound numbers for non-horizontal impacts (after Day & Goudie 1977).

	Recorded rebound values						*Hammer orientation*
	10	*20*	*30*	*40*	*50*	*60*	
		−5.4	−4.7	−3.9	−3.1	−2.3	
		−3.5	−3.1	−2.6	−2.1	−1.6	
Correction							
	+2.4	+2.5	+2.3	+2.0	+1.6	+1.3	
	+3.2	+3.4	+3.1	+2.7	+2.2	+1.7	

where R is the rebound number and δ is the compressive strength in pascals.

For purely comparative purposes it is unnecessary to convert R values to compressive strength. The rock-mass rating scheme of Selby (1980), for example, uses the raw R values.

The point load test

The point load test may be carried out in the field on specimens that are either cores or irregular lumps of rock. It is recommended that the test be carried out on cores only, that the load be applied across the diameter and that the minimum length of the core should be 1.5 times the diameter of the core (Bieniawski 1975, Broch & Franklin 1972). Tests conducted on irregular lumps and where the load is applied axially along a core may be subject to severe shape effects and give unpredictable and unreliable results. If either of these methods is used, larger numbers of tests should be made to allow for the scatter of results. The length of the loaded axis should not be less than 42 mm for any shaped specimen.

In order to judge whether the results from a particular point load test are valid, the fractured pieces of the specimen should be examined, and if a clean fracture runs from one loading point indentation to the other the test results can be accepted. If the fracture runs across some other plane, or if the points sink into the rock surface, causing both indentation and crushing, the results should be rejected.

A reasonable correlation exists between the point load index and the unconfined compressive strength, and details and tables for this conversion are given in Broch and Franklin (1972).

6.4   Soil description

There are three elements in the description of soils: on the largest scale a description of the site is needed, on the intermediate scale comes the description of the profile, and finally at the smallest scale comes the description of each individual horizon within the profile.

A site description should include a **locality name** and an appropriate **map reference**, the **elevation** and notes on the **regional and local relief**. The last term applies to the ground within a radius of c. 100 m and should include such things as the relation of the site to the main drainage lines, the slope angle and the aspect and form of the slopes. **Microrelief** – i.e. features individually less than c. 100 m² in area and with height differences of no more than a few metres – should also be described. The **weather** prior to and at the time of making the description should be noted in that it will affect observations on the soil-water status. Any evidence of **flooding** should be noted; so too should any evidence of **soil erosion or deposition** together with its intensity, age and whether by water, wind or mass movement. The area of ground occupied by **rock outcrops** should be recorded and any exposures described by reference to Section 6.1. The presence and frequency of **large stones or boulders** should also be noted. **Land use** and **local vegetation** should be described in some detail. For enclosed farmland specify the crop and the state of growth, and for semi-natural vegetation such as woodland, parkland and unenclosed farmland record the general type of plant community and the system of management. State the **form of the soil surface**, whether furrowed, moulded or ploughed, and the date of last ploughing if known. The **soil surface condition**, the state of aggregation of the surface (slaked, unslaked, capped) and **surface stoniness** should be noted. The lithology, size, shape and abundance of surface stones should be described using the terms recommended on pages 96–7.

The profile description should include a dimensioned **sketch** of the pit or cutting showing the disposition of the horizons. **Vertical and lateral relationships** between individual horizons should be clearly shown, the **maximum and minimum depths to the lower boundary** should be recorded and a note made of the **persistence of horizons**. Horizons should be named after complete description. **Samples** taken from the profile should be numbered consecutively from the surface downwards and their locations marked on the profile sketch. A note should also be made as to whether the pit or cutting is natural or artificial.

Each horizon within the soil profile should be fully described in terms of the features outlined in the following paragraphs. Care must be taken to ensure that the descriptions are as complete as possible and that any unusual features not included in the following list are recorded.

Soil texture can be fully described only after laboratory analysis has been carried out and then an appropriate textural name can be selected from either the classification used by the Soil Survey of England and Wales (Hodgson 1974) or that of the US Department of Agriculture (1951). A simple set of tests for determining the class of soil texture in the field based on Shaw (1928) is given below.

(a) **Sand** Loose and single-grained. Individual grains can be readily seen and felt. When squeezed dry it will fall apart when pressure is released. When moist it will form a cast which crumbles when touched.

(b) **Sandy loam** A sandy loam contains much sand but has enough silt and clay to make it somewhat coherent. Individual sand grains can be seen and felt. Squeezed when dry, it will form a cast which readily falls apart but when moist a cast can be formed which will bear careful handling without breaking.

(c) **Loam**   An even mixture of different grades of sand, silt and clay. Has a somewhat gritty feel, yet is fairly smooth and slightly plastic. Squeezed when dry, it forms a cast which will bear careful handling; the cast formed by squeezing when moist can be handled quite freely without breaking.

(d) **Silt loam**   A soil with a moderate amount of fine grades of sand and only a small amount of clay, over half of the particles being silt grade. When dry it appears cloddy, but the lumps can be readily broken and when pulverised it feels soft and floury. When wet, the soil readily runs together and puddles; either dry or moist it forms casts that can be freely handled without breaking. When moist and squeezed between thumb and forefinger, it will not form a ribbon, but presents a broken appearance.

(e) **Clay loam**   A fine textured soil which breaks into clods or lumps that are hard when dry. When moist soil is pinched between thumb and forefinger, it forms a thin ribbon which breaks readily, barely sustaining its own weight. The moist soil is plastic and will form a cast which will bear much handling. When kneaded in the hand it does not crumble readily but tends to form a heavy compact mass.

(f) **Clay**   A fine textured soil which usually forms very hard lumps or clods when dry and is quite plastic and usually sticky when wet. When moist soil is pinched between thumb and forefinger, it forms a long flexible ribbon. Some fine clays that are very high in colloid content are friable and lack plasticity in all moisture conditions.

Stoniness should be described in terms of the size, shape and roundness of individual stones using terms given in Section 6.5. The lithology of individual stones should be described by reference to Section 6.1 and Table 6.22 gives terms for the description of stone abundance. Charts for the visual estimation of abundance are given in Chapter 13.

Soil colours should be recorded by comparison with a Munsell Soil Colour Chart, a suitable alternative colour chart or in the absence of any chart the terms for the description of rock colour given in Table 6.6 may be used. When using a colour chart the 'moist soil colour', 'ped face colour', 'air-dry colour' and the 'rubbed colour' should all be recorded.

Mottling should be described in terms of the size, sharpness of boundaries and contrast of individual mottles. Size should be measured along the greatest dimension except in the case of linear forms which should be described in detail. The degree of contrast should be classed as faint,

Table 6.22 Terms for stone abundance (Hodgson 1974).

Term	Abundance (%)	Alternative terms for use where standard terms are awkward in context
stoneless	<1	
very slightly stony	1–5	few stones
slightly stony	6–15	common stones
moderately stony	16–35	many stones
very stony	36–70	abundant stones
extremely stony	>70	extremely abundant stones

distinct or prominent and the following terms should be used to describe the sharpness of boundaries:

sharp   a knife-edge boundary
clear   colour transition in space less than 2 mm
diffuse colour change over a distance of more than 2 mm.

Mottle abundance should be described using the terms below and the charts for estimating percentage cover given in Chapter 13.

none
few   occupy <2% of the surface
common   occupy 2–20% of the surface
many   occupy 20–40% of the surface
very many   occupy >40% of the surface.

The **distribution of organic matter** in the horizon should be described. Attention should be paid to whether it is intimately associated with the inorganic material, whether it is evenly or unevenly mixed, whether it occurs as coatings on mineral grains or as pellets. Organic matter lining or filling worm channels, pores or any other voids should be noted, as also should live or dead roots.

The **soil structure** should be described in terms of the shape, size and degree of development of the structural aggregates (peds, clods) and their spatial arrangement. The nature of the intervening pores and fissures should also be described. Peds are natural, relatively permament aggregates that persist through cycles of wetting and drying. Fragments or clods are less permanent aggregates formed at or near the surface by either frost action or cultivation. Table 6.23 shows illustrations of the various types of ped and gives a scheme for their nomenclature. The degree of ped development is assessed in the field by the ease with which the soil separates into peds and their durability when separated. This varies with the soil–water state and is normally described at the soil–water state in which the soil is found. Therefore, it is important that the soil–water state is recorded. The following terms may be used to describe the degree of ped development:

apedal	no observable aggregation: 'single-grained' – soil separates into primary particles 'massive' – breaks into masses which are easily broken
weakly developed	poorly formed, indistinct, weakly coherent peds that are barely observable in place
moderately developed	well formed moderately durable peds that are evident but not distinct in an undisturbed soil
strongly developed	durable peds that are quite evident in undisturbed soil.

Very strongly and very weakly developed classes may also be distinguished if required.

The direction, pattern, abundance and degree of development of fissures are closely related to the characteristics of the peds and can usually be inferred from the description of the peds although supplementary notes may prove useful.

Soil consistence is described under the headings of strength, maximum stickiness, cementation and plasticity. Since all of these properties are dependent on the soil–water state it is important that this be recorded.

The **strength** of the soil is taken as the resistance to crushing of an unconfined cube of soil *c*. 30 mm across and is normally assessed as strength in the vertical plane. Tables 6.24 and 25 give strength classes, conditions and characteristics of failure.

To assess **maximum stickiness** a crushed sample of soil is worked in the hand, and water is added or removed until the material is at its most sticky. Stickiness is then recorded using the following terms:

non-sticky	after release of pressure no material adheres
slightly sticky	soil material adheres perceptibly to the thumb and forefinger but is not stretched as they are separated
moderately sticky	adheres to the fingers but tends to stretch as they are parted
very sticky	soil adheres strongly to the fingers and is stretched as they are parted leaving part of the soil on each finger.

Table 6.23 Terms for the size and shape of peds and fragments (based on Hodgson 1974 and US Department of Agriculture 1951).

Size	Shape				
	Platy	Prismatic	Angular blocky	Sub-angular blocky	Granular
fine	fine platy <2 mm	fine prismatic <20 mm	fine angular blocky <10 mm	fine sub-angular blocky <10 mm	fine granular <2 mm
medium	medium platy 2–5 mm	medium prismatic 20–50 mm	medium angular blocky 10–20 mm	medium sub-angular blocky 10–20 mm	medium granular 2–5 mm
coarse	coarse platy 5–10 mm	coarse prismatic 50–100 mm	coarse angular blocky 20–50 mm	coarse sub-angular blocky 20–50 mm	coarse granular 5–10 mm
very coarse	very coarse platy >10 mm	very coarse prismatic >100 mm	very coarse angular blocky > 50 mm	very coarse sub-angular blocky >50 mm	very coarse granular >10 mm
illustration					

Table 6.24 Classification of soil strength (based on Hodgson 1974).

Strength classes	Force needed for failure in newtons (N)	Method used	Conditions of failure of 30 mm specimen cube or ped
loose	0		cube of soil cannot be obtained; this class is not used for ped strength of soil in a wet moisture state
very weak	<8	force applied between extended forefinger and thumb on horizontal faces of cube as it was oriented in the profile	cube or ped can be obtained but fails under gentle force
moderately weak	8–20		fails under gentle force
moderately firm	20–40		fails under firm force; force needed is significantly less than maximum force most people can exert slowly
very firm	40–80		fails under a force of 80 N, the maximum that most people can exert between extended thumb and forefinger
moderately strong	80–160	force applied slowly under foot on a hard flat surface or between both hands	fails with gentle force under foot; can be crushed between hands of average person
very strong	160–800	force applied slowly under foot on a hard flat surface	fails under the force which is applied by the full body weight of a man weighing c. 80 kg applied slowly
rigid	>800		withstands the force applied slowly under foot by a man of average weight

Cementation is assessed by placing an air-dry 30 mm cube of the soil in water for 1 h and then testing as outlined in Table 6.26.

Maximum plasticity is assessed by working a sample in the hand and adding water as required to achieve maximum plasticity. The soil is then formed into a roll c. 40 mm long and tested as outlined in Table 6.27.

Details of **roots and other plant remains** should be recorded. The size of roots should be described using the following terms:

very fine <1 mm
fine 1–2 mm
medium 2–5 mm
coarse >5 mm.

Table 6.25 Characteristics of failure of moist or wet soil specimens (Hodgson 1974).

Test	Result	Class
under gradually increasing pressure between extended thumb and forefinger a 30 mm specimen or cube	retains its size and shape until it fractures abruptly into peds or fragments	brittle
	is compressed but cracks develop and it ruptures before compression to half its original thickness	semi-deformable
	can be compressed to half its original thickness without cracks or rupture	deformable
if a handful of soil is squeezed in the hand	none of the material flows through the fingers	deformable
	it tends to flow into spaces between fingers but after exerting full pressure most of the soil is left in the hand	slightly fluid
	it flows easily through the fingers but a small residue is left in the hand after full pressure	moderately fluid
	it flows like a slightly viscous fluid between the fingers and very little or no residue is left after full pressure	very fluid

Table 6.26 Terms and tests for describing cementation (Hodgson 1974).

Term	Test result
uncemented	slakes when immersed in water for 1 h
very weakly cemented	can be crushed between extended forefinger and thumb (force <80 N)
weakly cemented	cannot be crushed between extended forefinger and thumb but fails when pressed under foot on a hard surface by a man of average weight (force 80–800 N)
strongly cemented	withstands weight of average man but breaks or crushes when struck by a blow of energy 3 J. This can be done by dropping an object (hammer, etc.) of known weight a distance of 0.3 m divided by the weight of the object in kilogrammes on to the specimen
very strongly cemented	unbroken when struck by a blow of 3 J

Table 6.28 gives terms for root abundance. Roots should be described as woody, fibrous or fleshy, and other living plant material such as bulbs, rhizomes, etc., should be described in general terms giving their average size and an indication of their distribution and abundance. Recognisable dead plant remains should be described in general terms and, in peaty

Table 6.27 Tests to determine maximum plasticity (Hodgson 1974).

Term	Test
non-plastic	a roll 40 mm long and 6 mm thick cannot be formed
slightly plastic	a roll 40 mm long and 6 mm thick can be formed and will support its own weight, but a roll 4 mm thick will not support its own weight
moderately plastic	a roll 40 mm long and 4 mm thick can be formed and will support its own weight, but a roll 2 mm thick will not support its own weight
very plastic	a roll 40 mm long and 2 mm thick can be formed and will support its own weight

Table 6.28 Descriptive terms for root abundance (Hodgson 1974).

Frequency class	Number of roots per 100 cm^2	
	Very fine and fine roots	Medium and coarse roots
few	1–10	1 or 2
common	10–25	2–5
many	25–200	>5
abundant	>200	—

Table 6.29 Field terms for calcium carbonate content (Hodgson 1974).

Term	$CaCO_3$ (%)	Audible effects	Visual effects
non-calcareous	<0.5	none	none
very slightly calcareous	0.5–1.0	just audible	none
slightly calcareous	1.0–2.0	moderately audible	slight effervescence just visible
moderately calcareous	2.0–5.0	distinctly audible	effervescence easily visible
calcareous	5.0–10.0	easily audible	moderate effervescence; bubbles up to 3 mm
very calcareous	>10	easily audible	general strong effervescence; large easily seen bubbles

horizons, the amount of humified material should be estimated and the peat classified as fibrous, semi-fibrous or amorphous.

The **calcium carbonate content** of the soil may be assessed by observing its reaction with a 10% hydrochloric acid solution and reference to Table 6.29.

Any **features of pedogenic origin** should be described. Notes should be made on any crystals, saline efflorescences, nodules or concretions present. These notes should include details of their abundance, shape, and composition and the presence, location, continuity and composition of any pans.

The distinctness and form of the **boundary to the next horizon** should be described using the following terms:

sharp	change occurs over a distance of < 5 mm
abrupt	change occurs over a distance of 5–25 mm
clear	change occurs over a distance of 25–60 mm
gradual	change occurs over a distance of 60–130 mm
diffuse	change occurs over a distance of > 130 mm
smooth	the boundary surface is plane with few irregularities
wavy	the boundary surface has broad shallow relatively regular pockets
irregular	the boundary surface has pockets that are deeper than they are wide
broken	the boundary is interrputed.

The **soil reaction** (pH) is best measured in the laboratory, but field determinations should be made with a portable meter and measured to one-tenth of a unit.

Nomenclature of soil horizons

Although the naming of soil horizons is partly dependent upon the observable field characteristics, complete and precise naming can only be carried out when the results of laboratory analyses are to hand. The nomenclature given here is purposely abbreviated and non-quantitative and is for preliminary field use only.

Litter layers
L   fresh litter deposited during the previous annual cycle

Organic horizons
F   partly decomposed litter remaining from earlier years in which some of the original plant structures are visible to the naked eye
H   well decomposed litter in which the original plant structures cannot be seen, often mixed with mineral matter
O   peaty horizons accumulated under wet conditions:

Of   fibrous peat
Om   semi-fibrous peat
Oh   amorphous peat
Op   mixed by cultivation

Mineral horizons
A   mineral horizon formed near the surface and characterised by incor-
poration of humified organic matter, disturbance by cultivation or
both:
Ah   uncultivated A horizon
Ap   mixed by cultivation
E   subsurface mineral horizon that is lighter in colour and contains less
organic matter than the immediately underlying horizon; it is
differentiated from the overlying horizons by a lighter colour and
lower organic content
Ea   without ferruginous mottles or nodules
Eb   dominant brownish colour due to evenly distributed free iron
oxide
B   mineral subsurface horizon without rock structure characterised by
either or both illuvial concentrations of silicate clay, iron, aluminium
or humus or by alteration of the original material involving the
solution and removal of carbonates
Bf   sharply defined horizon enriched in iron and carbon; black to
reddish-brown, brittle and less than 5 mm thick
Bh   B horizon containing translocated organic matter
Bs   B horizon characterised by enrichment with sesquioxides
Bt   B horizon characterised by translocated silicate clay occurring as
coats or as intra-ped concretions
Bw   horizon showing evidence of alteration by weathering, leaching
or structural reorganisation *in situ*
C   unconsolidated or weakly consolidated mineral horizon that retains
rock structure or otherwise lacks the properties of the overlying A, E
or B horizons
Cu   normally apedal or unstratified horizon that shows no evidence
of strong gleying
Cr   a weakly consolidated, little-altered substratum that is coherent
enough to prevent penetration by roots
Ck   horizon containing secondary $CaCO_3$ as concretions, coats or
nodules
Cy   horizon containing secondary accumulations of gypsum
R   hard or very hard bedrock that is continuous except for cracks.
Additional suffixes as below are frequently used:

G   intensely gleyed (reduced) horizon
g   mildly reduced horizon
m   continuously cemented horizon
x   horizon with fragipan characteristics.

6.5   Sediment and sedimentary rock description

As with soils there are three levels of description required. At the first level the description of the site should follow the same format as that suggested for soils at the beginning of Section 6.4.

The section or outcrop description should include precise **locational detail**, especially if the section in question is one of many closely spaced sections or part of a transect. The relation of the section to the **local landform** is particularly important in the case of sections in drift deposits. Note should be made of the **orientation of faces** examined and their relation to any adjacent section. Special note should be made of any possible **correlation between sections**. The **state of the exposure** should be recorded and some note should be made of the quality of the exposure, particularly if it is poor. Temporary or rapidly degrading sections should be identified in case any follow-up work is needed. Most of this information is best included on a detailed **field sketch** which should also show the disposition of the **boundaries between units** and any areas of non-exposure or poor quality exposure. A systematic method of **sample numbering** should be adopted, such as numbering from the top of the section with suffixes to denote multiple samples, and all sample spots should be shown on the field sketch.

Each unit or lithological unit should be fully described in terms of the characteristics outlined in the following paragraphs. Care should be taken to ensure that unusual features not covered in the sections that follow are recorded. In the case of the description of superficial sediments and weathered rocks the terms given for the description of soils may be useful.

The **grain size** of individual grains, clasts or well sorted sediments should be expressed using the appropriate terms from one of the accepted grade scales given in Table 6.30. For sediments or sedimentary rocks that are poorly sorted Figure 6.2 may be used to select an appropriate descriptive term. In the case of materials that contain a significant proportion of gravel grade material Figure 6.3 gives an appropriate terminology. The percentage of each constituent may be estimated in the field using Figure 13.2. The degree of **sorting** may be estimated in the field by reference to Figure 6.4.

Mechanical properties should be described where appropriate in terms

Table 6.30 Particle size grade scales.†

mm	φ	Wentworth (1922)	US Bureau of Soils	ASTM (1935)	BS 1377 (1967)
2048	−11				
1024	−10	boulders			
512	−9				
256	−8	--------			cobbles
128	−7	cobbles			
64	−6	--------		--- 76.1 --	
32	−5	pebbles	gravel	c	--- 60 --- c
16	−4			gravel − 19.0 -- f	-- 20 --- m
8	−3				gravel − 6 ---
4	−2	--------		--- 4.76 -- c	f
2	−1	granules		− 2.00 -- m	--- 2 --- c
1	0	vc			
0.5	1	sand c	sand c	sand − 0.42 --	sand − 0.6 ---
0.25	2	m	m		m 0.2--
0.125	3	f	f	f	f
0.0625	4	vf	− 0.1 − vf	-- 0.076 --	--- 0.06 --
0.0313	5	--------	--- 0.05 −		
0.0156	6	silt	silt	silt	silt c − 0.02 --
0.0078	7				m
0.0039	8		-- 0.005 −	-- 0.005 ---	− 0.006 ---
0.0020	9				f -- 0.002 ---
0.00098	10	clay	clay	clay	clay
0.00049	11				
0.00024	12				

†vc = very coarse; c = coarse, m = medium, f = fine, vf = very fine.

of strength, stickiness, cementation and plasticity. **Strength** may be assessed in qualitative terms using the tests and terms given in Table 6.31. At the upper end of the strength scale the tests given in Table 6.8 or Table 6.17 may be more appropriate. Alternative tests for the strength of soils are given in Table 6.24. **Stickiness, cementation** and **plasticity** may be described by reference to pages 98, 100, Table 6.26 and Table 6.27 respectively. In addition the nature of any cement should be identified and

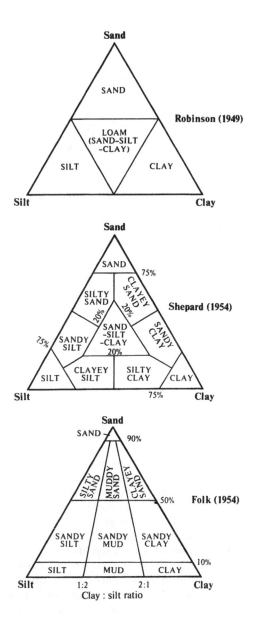

Figure 6.2 Nomenclature of sand, silt and clay mixtures.

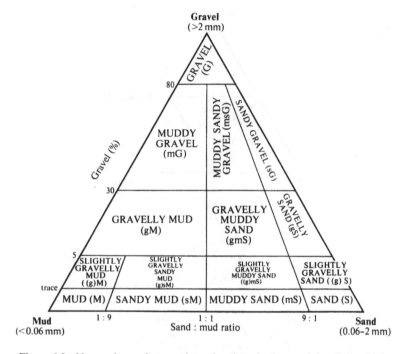

Figure 6.3 Nomenclature for gravel, sand and mud mixtures (after Folk 1954).

its distribution in relation to other lithological and structural properties ascertained. For geotechnical purposes samples should be taken for laboratory testing.

The **colour of sediments** should be assessed using a colour chart such as the Munsell Soil Colour Chart or the Geological Society of America Rock Colour Chart. If no colour chart is available the terms given in Table 6.6 may be used. Careful note should be made of the distribution of colour where it is uneven and the terminology for the description of mottling given on pages 96–7 may be useful in this context. Colour should be assessed for both fresh and weathered conditions.

The **type of constituent grains** should be recorded and where more than one type of grain is present an assessment of the relative abundance of the main types should be made. Accessory types and their abundance should also be noted. In the case of rock fragments, fossils and to a lesser extent heavy minerals, they should be identified as precisely as time and their size allows. In the case of fossils it is also useful to note their attitude,

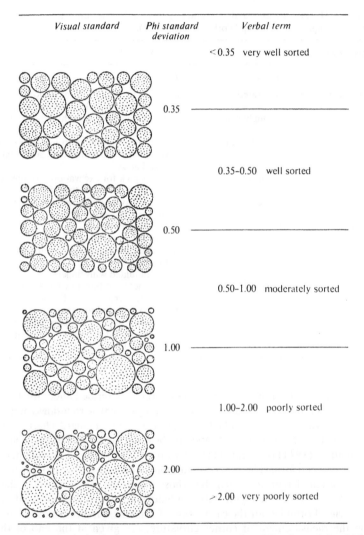

Visual standard	Phi standard deviation	Verbal term
		< 0.35 very well sorted
	0.35	
		0.35–0.50 well sorted
	0.50	
		0.50–1.00 moderately sorted
	1.00	
		1.00–2.00 poorly sorted
	2.00	
		> 2.00 very poorly sorted

Figure 6.4 Chart for the field estimation of sorting (modified from Folk 1968).

whether in life position, whether current orientated, and whether grouped or solitary. Figure 13.2 may be used to estimate the abundance of each constituent. Tables 6.1, 2, 3 & 4 may be used to assist in naming rock fragments and pebbles.

Table 6.31 Assessment and terminology for sediment strength (Geological Society Engineering Group Working Party 1977).

Sediment	Strength	
	Term	Definition
coarse-grained sediments	indurated	broken only with sharp pick blow, even when soaked; makes hammer ring
	strongly cemented	cannot be abraded with thumb or broken with hands
	weakly cemented	pick removes sediment in lumps, which can be abraded with thumb and broken with hands
	compact	requires pick for excavation; 50 mm peg hard to drive more than 50–100 mm
	loose	can be excavated with spade; 50 mm wooden peg easily driven
fine-grained sediments	hard	brittle or very tough
	stiff	cannot be moulded with fingers
	firm	moulded only by strong pressure of fingers
	soft	easily moulded with fingers
	very soft	exudes between fingers when squeezed
	friable	non-plastic, crumbles in fingers
peat	firm	fibres compressed together
	spongy	very compressible and open structure
	plastic	can be moulded in hands and smeared between fingers

The **form of the constituent particle** should be described under the headings of roundness, shape and surface markings. **Particle roundness** may be assessed visually in the field by comparison with standard charts. Two such charts are in common use and are shown in Figures 6.5 and 6. That of Krumbein (1941) is intended for use with pebbles of 16–32 mm diameter. In its reduced form (Fig. 6.5) comparability may suffer if used for particles substantially larger or smaller than those illustrated. The roundness chart of Powers (1953) in Figure 6.6 is intended for use only with sand-size particles. Formulae for the calculation of roundness indices, mostly based on the measurement of corner curvature, are given at the end of this chapter.

Shape, the relationship between the axial lengths, is best assessed by direct measurement and the calculation of shape indices such as those given at the end of this chapter. Figures 6.7 and 8 give two alternative classifications of particle shape.

The presence and type of any **surface markings** on individual grains

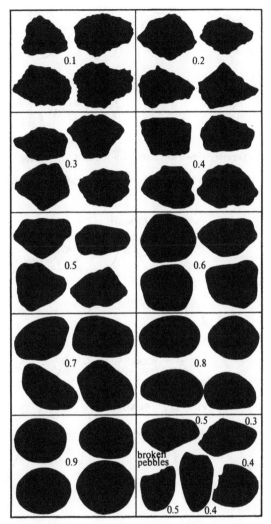

Figure 6.5 Images for the visual assessment of pebble roundness (based upon Krumbein 1941).

should be noted. The smaller grains should be examined with a hand lens. Note the presence of striations, frosting, polishing, chattermarks, gouges, etc. Note whether the features are common to all types of grain or

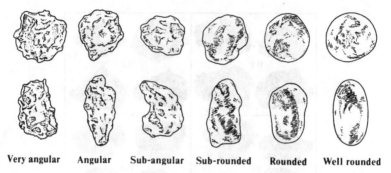

Very angular   Angular   Sub-angular   Sub-rounded   Rounded   Well rounded

Figure 6.6   Images for the visual assessment of sand grain roundness (based upon Powers 1953).

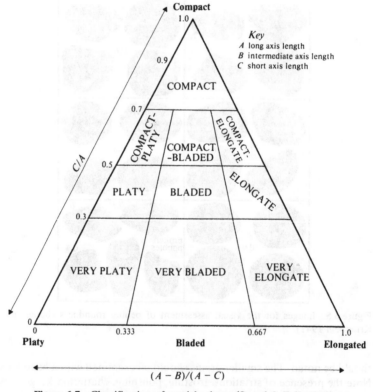

Figure 6.7   Classification of particle shape (Sneed & Folk 1958).

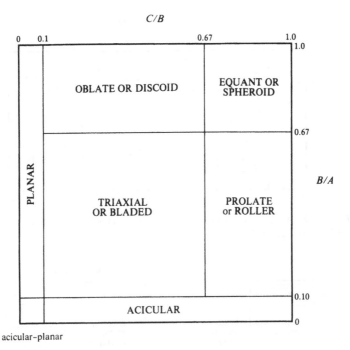

Figure 6.8 Classification of particle shape (based upon Zingg 1935, with additions of Brewer 1964). *A*, *B* and *C* are defined on Figure 6.7.

whether they are confined only to specific types. Any corrosion or grain coatings should also be noted.

The presence of any **concretions** or **nodules** should be recorded, their shape and composition should be described and note made of any apparent concentration or localisation.

Weathering state should be described using the terms given in either Table 6.16 or Table 6.18, and the degree of alteration may also be described in the same terms. The description of weathering in unconsolidated sediments may take the form of a soil description and the terminology outlined in Section 6.4 may be appropriate.

The **external geometry** of each unit should be described in terms of the bedding thickness, the nature of the layering and the nature of the boundary. **Bedding thickness** may be described using the terms given in Figure 6.9. The terms given in Figure 6.10 may be used to describe the layering in purely non-genetic terms. Alternative terms that may be used if more

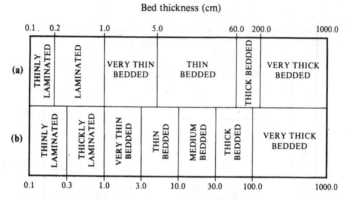

Figure 6.9 Alternative terminologies for bed thickness: (a) McKee and Weir (1953); (b) Ingram (1954).

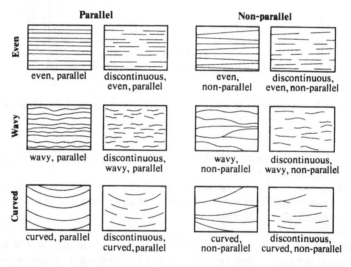

Figure 6.10 Descriptive terms for sedimentary layering (Campbell 1967).

appropriate are tabular, lenticular, linear, wedge or irregular. The **nature of the boundary** may be described using the terms given on page 103 for boundary distinctness and boundary form. When dealing with sediments as opposed to soils it may be useful to distinguish two types of gradational boundary:

mixed gradational   gradation occurs as a result of mixing of two distinct end members and a change in the relative abundance of each

continuously gradational gradation occurs as a continuous and progressive change from one type of material to the other without mixing of end members.

Additional terms that may be used to describe the form of boundaries include rippled, eroded, interdigitating, unconformable and conformable and should be used where they are more appropriate in context.

Note should be made of any pockets or **lenses of material of contrasting character** that may be included within the unit. Where appropriate the shape, size and precise nature of each such feature should be recorded.

The following checklist may be used to record the types of sedimentary structures that occur in each unit:

Inorganic structures

Fabrics	grain	clast	shell
Bedding types	cross bedding	rippled bedding	
	graded bedding	current bedding	lenses
	rhythmic bedding	fissility	flasers
Bedding plane markings	parting lineation	ripple marks	
	flute marks	tool marks	mud cracks
	scour marks	swash and rill marks	
Deformation structures	load casts	convolute bedding	clastic dikes
	channels	rip-up structures	folding
	faulting	frozen ground phenomena	
	flow structures	water escape structures	
Organic structures	burrows	tracks and trails	

Detailed descriptions and measurements of sedimentary structures are necessary when an environmental interpretation is required. In particular, cross-stratified and rippled units should be described using one of the schemes outlined below.

Three **classifications of cross-stratified units** are in common use. The schemes of Jopling and Walker (1967) and KcKee and Weir (1953) shown in Table 6.32 and Figure 6.11 respectively are easily used in the field but do not cater for all types or allow a precise genetic interpretation. Further subdivision of the McKee and Weir scheme may be achieved by using the following qualifying terms:

Shape of set	lenticular	set bounded by converging surfaces, at least one of which is curved
	tabular	set bounded by planar, essentially parallel surfaces
	wedge	set bounded by planar converging surfaces
Attitude	plunging or non-plunging	
Symmetry	symmetric	cross strata on opposite sides of the axial plane correspond in size and shape
	asymmetric	
Curvature of cross strata		convex, concave or straight
Inclination	high angle	20° or more
	low angle	less than 20°
Magnitude	small scale	cross strata less than 30 cm in length
	medium scale	cross strata 30 cm–6 m in length
	large scale	cross strata more than 6 m in length.

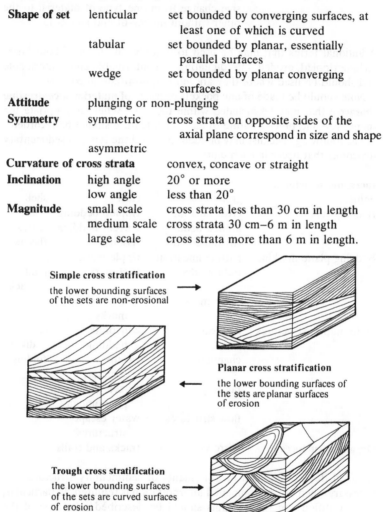

Simple cross stratification

the lower bounding surfaces of the sets are non-erosional

Planar cross stratification

the lower bounding surfaces of the sets are planar surfaces of erosion

Trough cross stratification

the lower bounding surfaces of the sets are curved surfaces of erosion

Figure 6.11 Terminology and classification of cross stratification (McKee & Weir 1953).

Table 6.32 Classification of cross lamination (Jopling & Walker 1967). Note that types A, B and S are a transitional series.

Feature	Type A	Type B	Type C	Type S
laminae	preserved on lee side only	continuous from stoss side to lee side	continuous from stoss side to lee side	continuous from stoss side to lee side
sediment	sand and silt, with some fines near base of slip face	sand and silt, fines concentrated near base of slip face and in trough	sand and silt restricted to stoss side, grading laterally into fines in the troughs	no selective concentration of sand, silt, or fines on particular parts of the ripples
ripple morphology	asymmetrical, amplitude variable	asymmetrical, amplitude variable	asymmetrical, amplitude decreases upward through coset	symmetrical, sinusoidal profile
grading of coset	ungraded	ungraded	graded, from fine sand and silt at the base to silt and fines at the top	ungraded
ratio of suspended to traction load	low and steady	intermediate and steady	intermediate and steady	high and steady
current strength stability	steady	steady	decreasing	decreasing

The classification proposed by Allen (1963) is more comprehensive but less easy to use in the field. To make field use of this scheme easier, diagrams illustrating the basic descriptive terms are shown in Figure 6.12. Using these diagrams a field description of the main features of each structure may be recorded. The key for identification in Table 6.33 may then be used to identify the type of cross stratification and a preliminary interpretation may then be made by reference to the genetic classification in Table 6.34. Illustrations of the main types in Allen's classification are shown in Figure 6.13.

Descriptions of **ripple marks** should be made using the terminology

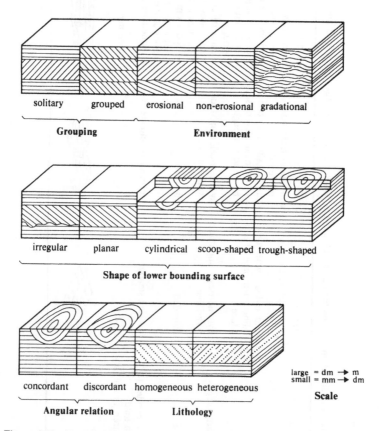

Figure 6.12   Descriptive terms for characteristics of cross stratification used in the classification of Allen (1963).

Table 6.33 Key for identification of cross-stratification type, in the classification of Allen (1963).

1	solitary or grouped?	(a)	solitary – go to 2 (large scale)
		(b)	grouped – go to 7
2	environment?	(a)	non-erosional – *ALPHA* (α)
		(b)	erosional – go to 3
3	lower bounding surface?	(a)	planar – go to 4
		(b)	irregular – *GAMMA* (γ)
		(c)	cylindrical – *ZETA* (ζ)
		(d)	scoop – *ETA* (η)
		(e)	trough – go to 5
4	lithology?	(a)	homogeneous – *BETA* (β)
		(b)	heterogeneous – *EPSILON* (ε)
5	angular relations?	(a)	concordant – *IOTA* (ι)
		(b)	discordant– *THETA* (θ)
6	large or small scale?	(a)	large – go to 7
		(b)	small – go to 8
7	environment?	(a)	erosional – go to 9
		(b)	non-erosional – *XI* (ξ)
8	environment?	(a)	erosional – go to 10
		(b)	gradational – go to 11
9	lower bounding surface?	(a)	planar – *OMIKRON* (o)
		(b)	scoop – *PI* (π)
10	lower bounding surface?	(a)	planar – *MU* (μ)
		(b)	scoop – *NU* (ν)
11	lower bounding surface?	(a)	irregular – *KAPPA* (κ)
		(b)	planar – *LAMBDA* (λ)

given in Figure 6.14 for the description of both the plan and elevation of the features. Useful information about the origin of ripple marks may be obtained by the calculation of various ripple indices. The measurements necessary are listed below and these allow calculation of those indices that Tanner (1967) cites as being those that best discriminate between ripples of different origins. Figure 6.15 shows the various combinations of ripple indices and may be used to discriminate between the various possible origins.

Measurements

Maximum ripple length in one set	L_{max}
Minimum ripple length in one set	L_{min}
Distance along one crest between two bifurcations	l_b
Crest length from end to end	C
Multiple measurements of ripple length	L_1, L_2, etc.
Distance parallel to crest along which curvature can be seen	l_d
Departure of crest line from the straight line in distance l_d	d
Horizontal length of stoss side	L_s

Table 6.34 Genetic classification of cross-stratified units (after Allen 1963).

	Group I	*Group II*	*Group III*
description	solitary sets, mostly large scale, bounded by planar or irregular surfaces	solitary sets, mostly large scale, filling hollows in the form of cylinders, scoops or troughs	grouped sets, either small, or large scale
types	alpha beta gamma epsilon xi	zeta eta theta iota	(A) small-scale sets: kappa lambda mu nu (b) large-scale sets: omikron pi
origin	due to migration of solitary banks with curving or linear fronts that in most cases are slip-off faces	due to cutting and filling of isolated channels, pits or hollows	due to migration of trains of different forms of small or large scale asymmetrical ripple marks, depending on the size and shape of the sets forming the coset

Horizontal length of lee side $\quad L_L$

Ripple height $\quad H$

Indices

Ripple index (RI)

$$= \frac{\text{ripple length}}{\text{ripple height}} = \frac{L}{H}$$

Ripple symmetry index (RSI)

$$= \frac{\text{length of horizontal projection of stoss side}}{\text{length of horizontal projection of lee side}} = \frac{L_s}{L_L}$$

Parallelism index no. 1 (PI_1)

$$= \frac{\text{length of curved part of crestline} \times \text{Minimum ripple length}}{\text{mean ripple length} \times \text{Maximum ripple length}} = \frac{l_d L_{min}}{L L_{max}}$$

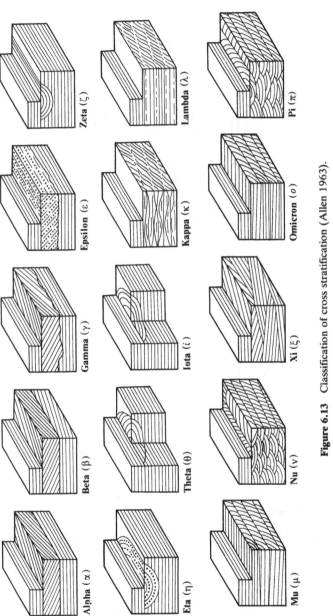

Figure 6.13 Classification of cross stratification (Allen 1963).

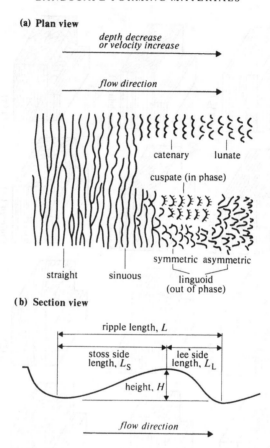

Figure 6.14 Terms for the plan and profile description of ripples (after Allen 1968).

Parallelism index no. 2 (PI$_2$)

$$= \frac{\text{maximum ripple length} - \text{minimum ripple length}}{\text{mean ripple length}} = \frac{L_{max} - L_{min}}{L}$$

Straightness index (SI)

$$= \frac{\text{length of curved part of crestline}}{\text{departure of curvature from straight (crest) line}} = \frac{l_d}{d}$$

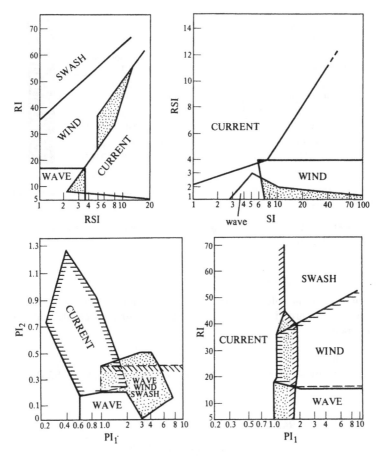

Figure 6.15 Relations between various combinations of ripple indices and environment of formation (after Tanner 1967).

Continuity index (CI)

$$= \frac{\text{crest length}}{\text{mean ripple length}} = \frac{C}{L}$$

Bifurcation index (BI)

$$= \frac{\text{distance between two bifurcations along the crest}}{\text{mean ripple length}} = \frac{l_b}{L}$$

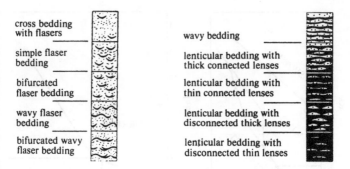

cross bedding with flasers

simple flaser bedding

bifurcated flaser bedding

wavy flaser bedding

bifurcated wavy flaser bedding

wavy bedding

lenticular bedding with thick connected lenses

lenticular bedding with thin connected lenses

lenticular bedding with disconnected thick lenses

lenticular bedding with disconnected thin lenses

Figure 6.16 Terms for the description of flaser and lenticular bedding (after Reineck & Wunderlich 1968).

Flaser and lenticular bedding may be described using the terms illustrated in Figure 6.16.

Particle form indices

Formulae for the calculation of the various commonly used parameters of particle form are given below. The notation is given separately for each parameter.

Sphericity

Indices of two-dimensional sphericity are given by Wadell (1935),

$$\varphi_w = d_c/D_c,$$

and by Riley (1941),

$$\varphi_r = (D_i/D_c)^{1/2},$$

where d_c = nominal section diameter, the diameter of a circle of area equal to that of the particle projection, D_c = diameter of the smallest circumscribing circle, D_i = diameter of the inscribed circle.

Wadell's and Riley's measures are based on measurements of the maximum cross-sectional area.

Indices of three-dimensional sphericity include two due to Krumbein (1941),

$$\psi_i = (BC/A^2)^{1/3},$$

$$\psi_k = \left\{ \left(\frac{B}{A}\right)^2 \frac{C}{B} \right\}^{1/3}$$

(the latter being an alternative form which may be obtained directly from the axial ratios used in the shape classification of Zingg (1935)), and a third due to Sneed and Folk (1958),

$$\psi_p = (C^2/AB)^{1/2},$$

where A = long axis length, B = intermediate axis length, C = short axis length.

Flatness

The roundness ratio is due to Wentwórth (1919):

$$F_{wl} = (A + B)/2C.$$

Flatness ratios are those of Wentworth (1919),

$$F_{w2} = (A + B + C)/3,$$

and Cailleux (1947),

$$F_c = \frac{A + B}{2C} \times 100,$$

where A = long axis length, B = intermediate axis length, C = short axis length.

Elongation

An index of two-dimensional elongation was given by Dapples and Rominger (1945),

$$E_p = W_p/L_p,$$

and an index of three-dimensional elongation by Scheiderhohn (1954),

$$E_s = B/A,$$

where W_p = width of particle projection, L_p = length of particle projection, A = long axis length, B = intermediate axis length.

Roundness

Indices of roundness include the roundness ratio of Wentworth (1933),

$$P_r = r_s/D,$$

the rounding index of Cailleux (1947),

$$P_i = 2r_s/A,$$

the degree of roundness of Wadell (1933),

$$P_d = \sum \frac{r}{R} \bigg/ N,$$

and the index of roundness of Ouma (1967),

$$X = 3\bar{r}/(A + B + C),$$

where r_s = radius of smallest corner, r = radius of each corner, $\bar{r}$ = average radius of all corners, R = radius of largest inscribed circle, D = mean grain radius = $(A + B)/4$ where A is the long axis length and B is the intermediate axis length, N = number of corners on the particle.

Note that the measures of Wadell and Ouma require that all corners are measured.

Miscellaneous shape indices

Miscellaneous shape indices include those due to Fleming (1965),

approximate or true nominal diameter, $D_e = (ABC)^{1/3}$;

Folk (1974),

volume, $V = (\pi/6)ABC$;

maximum projection area, MPA $= (\pi/4)AB$;

and Rivière and Ville (1967),

Morphological index, $K = \dfrac{1}{3}\left(\dfrac{A}{C} + \dfrac{2r}{C} + \dfrac{B}{C}\right)$;

where A = long axis length, B = intermediate axis length, C = short axis length, r = radius of smallest corner.

7 Fluvial processes

This chapter provides material to aid in measurement of fluvial processes and preliminary analysis and evaluation of field observations. Many of the relevant techniques are described in Goudie (1981), pp. 196–222. The first four sections concern discharge of water and the other three concern aspects of sediment in channels. Section 7.1 gives general recommendations concerning the velocity–area method and guidance for particular methods of velocity determination. Section 7.2 gives techniques by which resistance coefficients may be estimated in the field; the use of these coefficients is described in Section 7.1. Salt dilution gauging is considered in Section 7.3 and the use of gauging structures in Section 7.4. Section 7.5 is concerned with methods for estimation of sediment transport rates. The next section concerns relationships between bedforms, discharge and flow regime and the final section (7.7) is concerned with conductivity measurements. A table of constants for use in hydraulic calculations is also included (Table 7.6).

7.1 The velocity–area method

The recommended procedures for this method, based upon BS 3680 (1980), are outlined below. These are followed by material relevant to specific methods of velocity measurement. Finally the use of the velocity–area method for estimation of peak discharges is outlined.

Recommended procedures for velocity–area determination of discharge

In selecting a gauging site it is essential that:

(a) The channel reach should be straight and of uniform cross section and slope. When length is restricted the straight length upstream of the measurement section should be twice that downstream for current-meter measurements.
(b) Depth should be sufficient for effective immersion of the current-meter or floats.

(c) The channel should be clear and unobstructed by trees or other obstacles.

Additional points to be considered are as follows:

(a) The degree of accessibility of the site.
(b) The bed should not be subject to changes during the period of measurement. If measurements are to be repeated on further occasions, the bed and banks should be stable.
(c) All discharges should be contained within a definite channel or channels having substantially stable boundaries with well defined geometric dimensions.
(d) The site should be remote from tributary confluences, bends and natural or artificial obstructions if these are likely to cause disturbance to the flow.
(e) The channel should be free of aquatic growth.
(f) Sites at which vortices, backward flow or deadwater zones tend to develop should be avoided.
(g) Orientation of the reach should be such that the flow is as closely as possible normal to the prevailing wind.

When it is impossible to choose a site that satisfies all of the criteria, the undesirable characteristics should be minimised, but avoiding sites subject to variable backwater caused by constricted channel cross sections downstream. Flood plains, if unavoidable, should be of minimum width, as smooth as possible, without a distinct channel, and clear of bushes and trees. When it is necessary to make measurements near a bridge it is preferable to make them upstream, except that in situations where accumulations of logs, ice or other debris are likely to occur a downstream site should be used. For streams subject to the seasonal development of ice cover these additional considerations apply:

(a) The ice should be strong enough to bear the weight of persons and equipment.
(b) The site should be chosen so that the presence of slush or frazil ice is minimised. It is preferable to choose a site upstream rather than downstream of reaches of open water.
(c) Sites having many layers of ice should be avoided.

The numbers of verticals employed depends upon the width of the stream, but in general, for any but the smallest streams, a minimum of twenty is recommended, these being approximately equally spaced across the channel but taking into account characteristic points of the profile. This gives an error of less than 2% under normal conditions. Flow through the section between each pair of verticals should not be greater than 5% of the total.

Cross-sectional area should be defined by depth measurements at a series of closely spaced verticals, their position and spacing being chosen so as to define the profile accurately. The error in discharge determination can be reduced if knowledge of the continuous bed profile is used in constructing a cross section which is then used for calculation of the cross-sectional area between verticals. In general, intervals of not greater than one-fifteenth of the width should be employed for regular bed profiles, and intervals for irregular bed profiles should be not greater than one-twentieth of the width. These limits may, however, be relaxed for very small channels with regular profiles. Wherever possible the same verticals should be used for depth and velocity determination.

Velocity determination by current meter

When using point measurements the mean velocity should be determined according to the guidelines in Table 7.1 (BS 3680 1973). Wherever possible, two or more measurements should be employed within each vertical. Two points give an error of c. 5% in the determination of the mean velocity, whereas five give only c. 3% (Jansen 1979). Thirty seconds are sufficient to obtain a reasonable estimate of the true mean velocity at a vertical except at very low velocities, when 60 s are required. When more than four points are employed in each vertical this time can be shortened, but with a minimum of 10 s.

In the integration method the speed of lowering and raising the current meter should be less than 5% of the mean velocity of flow in the cross section and should in any case be less than 0.04 m s^{-1}. A correction is needed to allow for the part of the vertical not sampled near the bed of the stream. The height of this depends upon the type of current meter, usually being 0.15–0.25 m for larger streams. The corrections given in Figure 7.1 (Jansen 1979) are in terms of the total number of revolutions to be added.

Velocity determination by floats

The minimum duration of float movement should be 20 s, and measurement should be carried out in a straight reach of minimum length 100 m. The averaged velocity from a series of measurements must be multiplied by a coefficient to convert it to mean velocity in the vertical. This coeffi-

Table 7.1 Averaging procedures and applicability of current metering methods. D = depth; v_n = velocity at the fraction n of the depth (D) from the surface; $k_1 \approx 0.95$, but should be verified by using the velocity-distribution method; k_2 = in general 0.84–0.90, with 0.88–0.90 being characteristic of smooth beds. More extreme values of k_2 may occur and it is advisable to compute a coefficient specific to each section and stage by correlating 'surface' velocity with either velocity at $0.6D$ or mean velocity from the integration method.

Number of points in vertical	Depths of measurement points (D = depth)	Averaging procedure	Application
1	$0.6D$	$\bar{v} = v_{0.6}$	D small
1	$0.5D$	$\bar{v} = k_1 v_{0.5}$	D small
1	immediately below surface	$\bar{v} = k_2 v_{\text{surface}}$	'flashy' sections
1	$0.5D$	$\bar{v} = 0.88 v_{0.5}$	ice-covered $D < 1$ m
1	$0.6D$	$\bar{v} = 0.92 v_{0.6}$	ice-covered $D < 0.5$ m
2	$0.2D$ and $0.8D$	$\bar{v} = 0.5(v_{0.2} + v_{0.8})$	$D > 0.5$ m (0.3 m if miniature meter used)
3	$0.2D$, $0.6D$ and $0.8D$	$\bar{v} = 0.25(v_{0.2} + 2v_{0.6} + v_{0.8})$	$D > 0.5$ m
3	$0.15D$, $0.5D$ and $0.85D$	$\bar{v} = (v_{0.15} + v_{0.5} + v_{0.85})/3$	ice-covered
5	$0.2D$, $0.6D$, $0.8D$, surface and bed	$\bar{v} = 0.1(v_{\text{surface}} + 3v_{0.2} + 3v_{0.6} + 2v_{0.8} + v_{\text{bed}})$	free from ice and aquatic growth
6	$0.2D$, $0.4D$, $0.6D$, $0.8D$, surface and bed	$v = 0.1(v_{\text{surface}} + 2v_{0.2} + 2v_{0.4} + 2v_{0.6} + 2v_{0.8} + v_{\text{bed}})$ or planimeter isovel cross section	much ice or vegetation

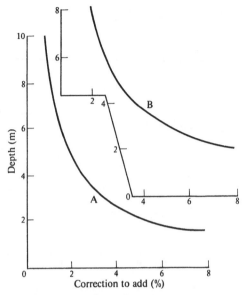

Figure 7.1 Correction to be added to the number of revolutions recorded in the integration method of velocity determination by current-meter, assuming a logarithmic vertical velocity distribution: curve A, unsampled part of the vertical = 0.15 m; curve B, unsampled part of the vertical = 0.2 m (based upon Jansen 1979).

cient generally lies in the range 0.80–1.0, with the value being chosen according to local conditions. For surface floats it is generally in the range 0.84–0.90, with the higher values being characteristic of smooth beds. For double floats, with the subsurface body at 0.6 of the depth (D) from the surface, the coefficient is 1.0 and with the subsurface body at $0.5D$ the coefficient is 0.96.

The mean velocity in a vertical may also be estimated from the observed velocity of a vertical rod float by multiplying this velocity by a coefficient read from Figure 7.2. The lines represent different values of P, which is defined in the figure. Mean velocity in the section ($\bar{v}$) for channels of rounded cross section may be estimated from the maximum surface velocity observed by use of floats (v'), from $\bar{v} = kv'$, where k is a coefficient that depends upon the depth (D) according to $k = 0.715D^{0.072}$. Further recommendations concerning the use of floats will be found in BS 3680 (1964, 1980).

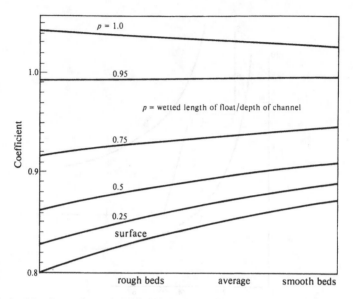

Figure 7.2 Correction coefficient for estimation of mean velocity in a vertical from observed velocity of vertical rod floats (based upon BS 3680 1964).

Velocity determination from formulae

Stream velocity may be determined from any of the so-called uniform-flow formulae. The most useful methods are given below.

The Manning equation

$$\bar{V} = R^{2/3}S^{1/2}/n ,$$

where $\bar{V}$ = mean channel velocity (m s^{-1}), R = hydraulic radius (m) = $A/P \approx$ depth for channels wide relative to their depths, A = channel cross-section area (m^2), P = channel cross-section wetted perimeter (m), S = water surface slope (as gradient, e.g. m m^{-1}), n = Manning's n resistance coefficient. Methods for estimating n are given in Section 7.2.

$R^{2/3}$ for given values of R are in Table 7.2, since this calculation of $R^{2/3}$ is not always readily executed on a pocket calculator. A nomogram for the solution of the Manning equation is given in Figure 7.3 after Chow (1959). To use this nomogram a line is laid from the value of hydraulic radius through that of Manning's n and its intersection with the pivot line

Table 7.2 One-sixth and two-thirds powers of numbers for use in hydraulic calculations.

R	$R^{1/6}$	$R^{2/3}$	R	$R^{1/6}$	$R^{2/3}$
.05	.6070	.1357	2.10	1.1316	1.6399
.10	.6813	.2154	2.15	1.1361	1.6658
.15	.7289	.2823	2.20	1.1404	1.6915
.20	.7647	.3420	2.25	1.1447	1.7171
.25	.7937	.3969	2.30	1.1489	1.7424
.30	.8182	.4481	2.35	1.1530	1.7676
.35	.8395	.4966	2.40	1.1571	1.7926
.40	.8584	.5429	2.45	1.1611	1.8174
.45	.8754	.5872	2.50	1.1650	1.8420
.50	.8909	.6300	2.55	1.1688	1.8665
.55	.9052	.6713	2.60	1.1726	1.8908
.60	.9184	.7114	2.65	1.1764	1.9150
.65	.9307	.7504	2.70	1.1800	1.9390
.70	.9423	.7884	2.75	1.1836	1.9629
.75	.9532	.8255	2.80	1.1872	1.9866
.80	.9635	.8618	2.85	1.1907	2.0102
.85	.9733	.8973	2.90	1.1942	2.0336
.90	.9826	.9322	2.95	1.1976	2.0569
.95	.9915	.9664	3.00	1.2009	2.0801
1.00	1.0000	1.0000	3.05	1.2042	2.1031
1.05	1.0082	1.0331	3.10	1.2075	2.1261
1.10	1.0160	1.0656	3.15	1.2107	2.1489
1.15	1.0236	1.0977	3.20	1.2139	2.1715
1.20	1.0309	1.1292	3.25	1.2171	2.1941
1.25	1.0379	1.1604	3.30	1.2202	2.2165
1.30	1.0447	1.1911	3.35	1.2232	2.2389
1.35	1.0513	1.2215	3.40	1.2263	2.2611
1.40	1.0577	1.2515	3.45	1.2292	2.2832
1.45	1.0639	1.2811	3.50	1.2322	2.3052
1.50	1.0699	1.3104	3.55	1.2351	2.3271
1.55	1.0758	1.3393	3.60	1.2380	2.3489
1.60	1.0815	1.3680	3.65	1.2408	2.3706
1.65	1.0870	1.3963	3.70	1.2437	2.3922
1.70	1.0925	1.4244	3.75	1.2464	2.4137
1.75	1.0978	1.4522	3.80	1.2492	2.4351
1.80	1.1029	1.4797	3.85	1.2519	2.4564
1.85	1.1080	1.5070	3.90	1.2546	2.4777
1.90	1.1129	1.5340	3.95	1.2573	2.4988
1.95	1.1177	1.5608	4.00	1.2599	2.5198
2.00	1.1225	1.5874	4.05	1.2625	2.5408

is marked. A second line is then laid from this point to the appropriate value of slope and the velocity is read off from the velocity axis.

The Chézy equation

$$V = C\,(R\,S)^{1/2},$$

where C = Chézy's C flow resistance factor. Methods for estimation of C are given in Section 7.2.

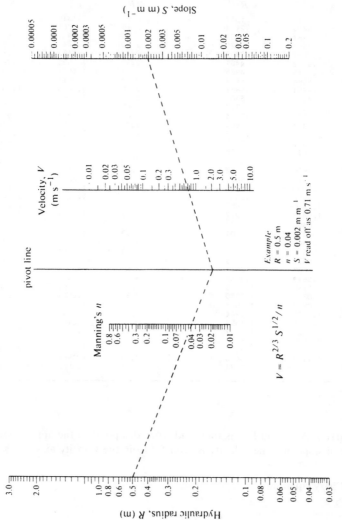

Figure 7.3 Nomogram for evaluation of the Manning equation (after Chow 1959). For explanation of use see text.

The Darcy–Weisbach equation

$$\bar{V}^2 = 8\,gRS/f\,,$$

where g = the gravitational constant (m s^{-2}), f = the Darcy–Weisbach resistance factor. f may be derived from the Colebrook–White equation (p. 138); the use of this equation to derive mean velocity is described below (pp. 138–9).

The Chézy and Manning resistance factors are related by $C = R^{1/6}/n$. Values of $R^{1/6}$ for given values of R are also included in Table 7.2.

Although valid in most applications, the above formula may be modified to the Pavlovskiĭ formula if required: $C = R^y/n$, where

$$y = 2.5\,n^{1/2} - 0.13 - 0.75\,R^{1/2}\,(n^{1/2} - 0.10)\,,$$

which is valid for $0.1 < R < 3.0$ m and $0.011 < n < 0.04$. This may be approximated by $y = 1.5\,n^{1/2}$ for $R < 1.0$ m and $y = 1.3\,n^{1/2}$ for $R > 1.0$ m.

The Darcy–Weisbach resistance factor (f) is related to C and n by $n = R^{1/6}\,(f/8g)^{1/2}$ and $C = (8g/f)^{1/2}$.

For fuller discussion of uniform-flow formulae see Chow (1959).

Estimation of peak discharge by the slope–area method

This method, after Benson (1968), should be used for estimation of flood discharges after the flood wave has passed. The reach employed should be:

(a) of uniform cross section
(b) free from obstructions, disturbances and backwater effects
(c) of length not less than 75 times the mean depth
(d) of length not less than 5 times the mean width
(e) greater than 300 m in length
(f) with a surface fall of not less than 0.15 m over the reach.

The cross section is divided into parts $(1, 2, \ldots, n)$ selected so that the Chézy resistance coefficient (C_j) within a part can be regarded as constant (Fig. 7.4). The cross-sectional area (A_j) and hydraulic radius (R_j) are determined for each part. Discharge through each is then estimated from

$$Q_j = \bar{V}A_j = C_jA_jR_j^{0.5}S^{0.5},$$

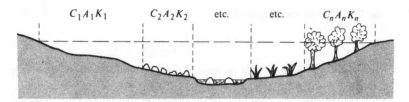

Figure 7.4 Division of cross section into internally homogenous parts in slope-area method of discharge estimation.

where Q_j = discharge of each part, $\bar{V}$ = mean velocity of each part, S = mean slope through the reach, $C_jA_jR_j^{0.5}$ is the conveyance (K_j) for each part of the cross section.

The discharge of the entire cross section is given by

$$Q = S^{0.5} \sum_{j=1}^{j=n} K_j.$$

If the cross sections differ from beginning to end of the reach, the geometric mean conveyance is used as

$$K_j = (K_{j1}K_{j2})^{0.5}.$$

7.2 Estimation of resistance coefficients in the field

Many methods are available for field estimation of resistance coefficients such as those described in the above previous section. Selection of the most appropriate method should be made on the basis of type and size of channel, according to the following system:

(a) For channels with coarse (i.e. larger than sand size) bed material the visual comparison method using photographs such as those reproduced in Barnes (1967) and Chow (1959) may be used. Photographs, however, are not reproduced here because of space limitations and if it is intended to use this method the original sources should be used. An alternative is to use formulae (given on p. 137) relating resistance to bed material particle size.

(b) For channels with sand beds the procedure of Simons and Richardson (1962a & b, 1971) may be used, as described on page 139.

(c) For all types of smaller channels the component method of Cowan (1965), described on pages 140–1, may be used.

(d) For all channels the descriptions of Chow (1959), given on page 141, may be used.

Although most of the methods are expressed in terms of determination of Manning's n, they may also be used to derive other resistance coefficients from the relationships described in Section 7.1. For a rigorous discussion of resistance coefficients and estimation methods see Simons and Şentürk (1976).

Estimation of resistance coefficients from coarse bed material particle size

Numerous equations have been proposed with a form similar to that of Strickler (1923):

$$n = kd^{1/6},$$

where n = Mannings's n, d = a measure of bed material particle size (m) in the form of a particular percentile value of the cumulative frequency distribution of particle sizes (e.g. d_{50} = the fiftieth percentile or median value), k = a coefficient.

Some commonly used values of k from Simons and Şentürk (1976) and Bray (1979) are

d	k	
d_{50}	0.0594	(d in m)
	0.0410	(d in m)
d_{65}	0.0132	(d in mm)
	0.0569	(d in m)
d_{90}	0.0385	(d in m)
	0.0523	(d in m)
d (uniform sand)	0.0474	(d in mm)

A simple variant of the general formula is.

$$n = 0.113d_{75}^{0.5}/D^{0.33}$$

where D = water depth (m).

A more sophisticated development (Limerinos 1970) employing Imperial units is

$$n = 0.0926R^{1/6}/\{a + 2.0 \log(R/d)\} ,$$

where R = hydraulic radius (ft), a = a coefficient whose value depends upon which axis of the bed material and percentile of the size distribution are used, according to the following relationship:

	Minimum diameter	Intermediate diameter
d_{84} (ft)	0.76	1.16
weighted particle size, p_w (ft)	0.50	0.90

where $p_w = 0.6d_{84} + 0.3d_{50} + 0.1d_{16}$.

The Darcy–Weisbach friction factor (f) may be estimated from the Colebrook–White equation (Hey 1979) in the form

$$1/f^{0.5} = 2.03 \log(aR')/3.5d_{84} ,$$

where d_{84} = the eighty-fourth percentile of the cumulative frequency distribution of the bedload's intermediate diameter, R' = an effective hydraulic radius allowing for cross-sectional variations in roughness height. R' is calculated from

$$R' = A/\{P_b + (Y_l/\sin \theta_l) + (Y_r/\sin \theta_r)\} ,$$

where A = cross-sectional area (m²), P_b = wetted perimeter of bed (m), Y_l and Y_r = perpendicular distances from left and right banks to the point of maximum velocity (m), θ_l and θ_r = left and right bank slope angles, a = a coefficient read from Figure 7.5.

Steps in the application of this procedure are therefore:

Step 1: Calculate R' from the cross section.
Step 2: Calculate R'/Y_b, as explained in Figure 7.5.
Step 3: Read off a from Figure 7.5.
Step 4: Calculate f from the first equation given.

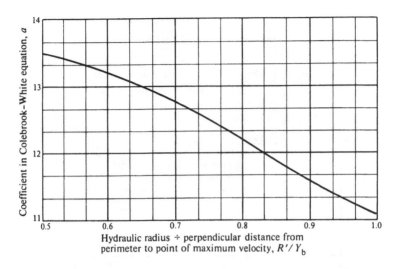

Figure 7.5 Derivation of coefficient a in Colebrook–White equation (Hey 1979). Y_b represents the maximum distance from the perimeter (usually the bed) to the point of maximum velocity.

Estimation of resistance coefficient for sand bed streams

The range of values within which resistance should lie may be estimated by the following procedure, after Simons and Richardson (1962a & b, 1971).

Step 1: Determine the median diameter of bed material.

Step 2: Determine the Froude number (F) from

$$F = V/(gD)^{1/2},$$

where V = water velocity (m s^{-1}), g = gravitational constant (m s^{-2}), D = water depth (m).

Step 3: Read off the limits of Manning's n for the appropriate size of sediment, value of F and associated bedforms from Figure 7.6.

Note that resistance declines with increasing discharge as bedforms become planed out; some values indicative of this are given in Figure 7.13.

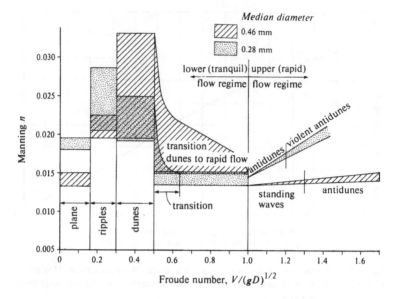

Figure 7.6 Variation of Manning's n with Froude number for sand bed streams (Simons & Richardson 1962a,b, 1971).

Component method for estimation of resistance

This method, of Cowan (1956), is unsuitable for large channels (of hydraulic radius greater than c. 5 m). In using it care must be taken to avoid re-evaluating or double-counting conditions under more than one component.

n is estimated from the formula

$$n = (n_0 + n_1 + n_2 + n_3 + n_4)m_5.$$

n_0 is a basic value for a straight uniform channel in the appropriate material. n_1 incorporates the effects of surface irregularities. n_2 incorporates the effects of variations in shape and size of the channel cross section. n_3 incorporates the effects of obstructions. n_4 incorporates the effects of vegetation. m_5 incorporates the effects of channel meandering (note that this is a multiplier).

Values of these components are as follows:

Material, n_0	
earth	0.020
rock	0.025
fine gravel	0.024
coarse gravel	0.028

Degree of surface irregularity, n_1

smooth	0.000
minor (e.g. only minor slumping)	0.005
moderate (e.g. moderate slumping)	0.010
severe (e.g. badly slumped, or	
irregular rock surfaces)	0.020

Variation of channel cross section, n_2

gradual	0.000
alternating occasionally	0.005
alternating frequently	0.010–0.015

Relative effect of obstructions (e.g. debris, roots, boulders), n_3

neglible	0.000
minor	0.010–0.015
appreciable	0.020–0.030
severe	0.040–0.060

Vegetation, n_4

none	0.000
low	0.005–0.010
medium	0.010–0.025
high	0.025–0.050
very high	0.050–0.100

Degree of meandering, m_5 (multiplier)

minor (sinuosity < 1.2)	1.00
appreciable (sinuosity 1.2–1.5)	1.15
severe (sinuosity > 1.5)	1.30

Determination of resistance by comparison with descriptions

The descriptions given in Tables 7.3 and 7.4, from Chow (1959), may be applied to natural channels. Minimum, normal and maximum values of Manning's *n* are given where appropriate. Table 7.3 applies to minor streams and flood plains. Table 7.4 applies to major streams and streams with coarse bed materials or rock-cut channels. Note that as the form of bed roughness changes in alluvial channels from dunes through transition to plane bed or standing waves the magnitude of *n* decreases by approximately 50% (see Fig. 7.13).

7.3 Discharge measurement by dilution gauging

The basic formulae required in this technique are presented here as an *aide memoire*. For full details and descriptions of the techniques see Church (1975), Jansen (1979) and Goudie (1981).

Constant rate injection

$$Q = \cdot TN ,$$

where Q = discharge, T = rate of injection, and $N = C_1/C_2$ where C_1 =

Table 7.3   Manning's n for minor streams and flood plains (after Chow 1959).

	Manning's n
MINOR STREAMS (bank full width <30 m)	
(a) *Streams on plain*	
(1) Clean, straight, full-stage, no rifts or deep pools	0.025–0.030–0.033
(2) As (1) but more stones and weeds	0.030–0.035–0.040
(3) Clean, winding, some pools and shoals	0.033–0.040–0.045
(4) As (3) but with some weeds and stones	0.035–0.045–0.050
(5) As (4) but lower stages, less efficient slopes and sections	0.040–0.048–0.055
(6) As (4) but with more stones	0.045–0.050–0.060
(7) Sluggish reaches, weedy, deep pools	0.050–0.070–0.080
(8) Very weedy reaches with deep pools	0.075–0.100–0.150
(9) Floodways with heavy stands of timber and vegetation	0.075–0.100–0.150
(b) *Mountain streams* with no vegetation in channel, usually with steep banks and bank vegetation submerged at high stages	
(1) Bed of gravel, cobbles and a few boulders	0.030–0.040–0.050
(2) Bed of cobbles with large boulders	0.040–0.050–0.070
FLOOD PLAINS	
(a) *Pasture, no brush*	
(1) Short grass	0.025–0.030–0.035
(2) Long grass	0.030–0.035–0.050
(b) *Cultivated areas*	
(1) No crops	0.020–0.030–0.040
(2) Mature row crops	0.025–0.035–0.045
(3) Mature field crops	0.030–0.040–0.050
(c) *Brush*	
(1) Scattered brush, heavy weeds	0.035–0.050–0.070
(2) Light brush and trees, in winter	0.035–0.050–0.060
(3) As (2), in summer	0.040–0.060–0.080
(4) Medium to dense brush, in winter	0.045–0.070–0.110
(5) As (4), in summer	0.070–0.100–0.160
(d) *Trees*	
(1) Dense straight willows, summer	0.110–0.150–0.200
(2) Cleared land with stumps, not sprouting	0.030–0.040–0.050
(3) As (2), but heavy sprout growth	0.050–0.060–0.080
(4) Heavy stands of timber, a few fallen trees, little undergrowth, flood stage below branches	0.080–0.100–0.120
(5) As (4), but floods reaching branches	0.100–0.120–0.160

concentration of tracer in injected solution and C_2 = concentration of tracer in sampled section.

If tracer is naturally present in the stream at concentration C_0, then $N = C_1/(C_2 - C_0)$.

Table 7.4 Values of Manning's n for major streams and streams with coarse bed materials or rock-cut channels (Chow 1959).

	Manning's n
MAJOR STREAMS (bank full width >30 m)	
(1) Regular sections with no boulders or brush	0.025–0.060
(2) Irregular and rough sections	0.035–0.100
CHANNELS WITH RELATIVELY COARSE BED MATERIALS AND ROCK-CUT CHANNELS	
gravel, 4–8 mm in diameter	0.019–0.020
gravel, 8–20 mm in diameter	0.020–0.022
gravel, 20–60 mm in diameter	0.022–0.027
pebbles and shingle, 60–110 mm in diameter	0.027–0.030
pebbles and shingle, 110–250 mm in diameter	0.030–0.035
smooth and uniform rock-cut channels	0.025–0.035–0.040
jagged and irregular rock-cut channels	0.035–0.040–0.050

Slug injection

$$Q = (C_1 - C_0)V \Big/ \int_0^t [(C_t - C_0)]\, dt,$$

where V = the volume of tracer injected, C_1 = the concentration of tracer injected, C_0 = the natural concentration of the tracer, C_t = the concentrations of tracer measured during the time taken for the wave of tracer to pass.

Downstream sampling distance

Complete mixing is achieved when the distance of the sampling point from the injection point (L) is greater than the value given by

$$L = 0.13\{(0.7C + b)/g\}(b^2/d)$$

where C = the Chézy resistance coefficient (see p. 133), b = average width of stream (m), d = average depth of stream (m), g = gravitational constant (m s^{-2}).

7.4 Gauging structures

The measurement of discharge by use of permanent gauging structures is beyond the scope of this manual. However, some basic formulae are included here for use when it is wished to establish temporary gauging stations during a limited field season. For fuller details of techniques and relationships reference should be made to Ackers *et al.* (1978) and BS 3680 (1965). Only sharp-edged weirs are considered because these are most appropriate for temporary locations.

V-notch weirs

Discharge (Q) is given by

$$Q = (8/15)(2g)^{1/2}C_d \tan(\theta/2)H^{5/2} ,$$

where g = gravitational constant (m s^{-2}), C_d = discharge coefficient (see Table 7.5), θ = the angle between the sides of the notch, H = the total head of water (m).

The limitations applying to this relationship are that

(a) H shall not be less than 0.05 m or greater than 0.38 m,
(b) the vertex height (P) shall exceed 0.45 m,
(c) H/P shall not exceed 0.4,
(d) the width of the approach channel (B) shall not exceed 0.9 m, and
(e) H/B shall not exceed 0.20.

Rectangular-notch weirs

Discharge is given by

$$Q = (2/3)(2g)^{1/2}C_d bh_g^{3/2} ,$$

where b = notch width, h_g = gauged head + 0.0012 m, and other terms are as for V-notch weirs. In this case C_d is given by

$$C_d = 0.602 + 0.083h/p ,$$

where h = the gauged head (m), p = the weir height (m).

The limitations applying to the rectangular weir relationship are that

(a) h/p shall not exceed 1.0,
(b) the head (h) shall be between 0.03 m and 0.75 m,
(c) the weir width (b) shall be at least 0.3 m, and
(d) the weir height shall not be less than 0.10 m.

Table 7.5 Discharge coefficients for V-notch weirs (after BS 3680 1965).

Head (m)	90° V-notch	Half 90° V-notch	Quarter 90° V-notch
0.05	0.6080	0.6153	0.6508
0.06	0.6032	0.6114	0.6417
0.07	0.5994	0.6084	0.6352
0.08	0.5964	0.6060	0.6298
0.09	0.5937	0.6040	0.6256
0.10	0.5917	0.6021	0.6219
0.11	0.5898	0.6005	0.6187
0.12	0.5885	0.5989	0.6162
0.13	0.5876	0.5976	0.6139
0.14	0.5868	0.5964	0.6119
0.15	0.5861	0.5955	0.6102
0.16	0.5857	0.5946	0.6085
0.17	0.5853	0.5938	0.6070
0.18	0.5851	0.5930	0.6057
0.19	0.5850	0.5923	0.6047
0.20	0.5849	0.5918	0.6038
0.21	0.5848	0.5913	0.6029
0.22	0.5847	0.5908	0.6021
0.23	0.5846	0.5904	0.6014
0.24	0.5846	0.5901	0.6008
0.25	0.5846	0.5898	0.6002
0.26	0.5846	0.5896	0.5997
0.27	0.5846	0.5892	0.5992
0.28	0.5847	0.5890	0.5988
0.29	0.5847	0.5888	0.5984
0.30	0.5848	0.5885	0.5980
0.31	0.5849	0.5882	0.5976
0.32	0.5850	0.5881	0.5972
0.33	0.5850	0.5880	0.5968
0.34	0.5851	0.5879	0.5964
0.35	0.5852	0.5877	0.5960
0.36	0.5853	0.5875	0.5956
0.37	0.5854	0.5874	0.5952
0.38	0.5855	0.5872	0.5948

7.5 Scour and sediment transport

Most relationships used in examination of sediment transport are too complex for ready evaluation in the field and they rarely yield estimates of sediment transport rates that accord closely with field observations. The material included here is therefore intended largely as a basis for preliminary estimates and evaluation of field data. For more detailed material reference should be made to Graf (1971) and Simons and Şentürk (1976).

Figure 7.7 is the curve of Hjulstrom (1935) showing the mean velocity necessary to initiate motion and that below which sedimentation occurs. This is strictly applicable only to sediments of a single size, but is often used to derive approximate values for sediment mixtures. A simple equation for the estimation of bedload transport rate is the Duboys equation (Graf 1971). The volume of bedload transport per unit width of stream per unit time (q_s) is

$$q_s = \chi \tau_0 \{ \tau_0 - (\tau_0)_{cr} \} ,$$

where χ = a characteristic sediment coefficient, τ_0 = shear stress = $\gamma DS (\text{kg m}^{-2})$, γ = unit weight of water (N m^{-3}), D = depth (m), S = channel slope (as gradient, e.g. m m^{-1}), $(\tau_0)_{cr}$ = shear stress at the critical condition when movement is just beginning.

τ_0 can be evaluated by direct measurement and calculation. $(\tau_0)_{cr}$ and χ can be read off Figure 7.8 (Zeller 1963) for the appropriate median sediment size.

Figure 7.9 is Shields' (1936) curve showing the relationship between Shields' entrainment function (F_s) and the shear Reynolds number (Re*) given by

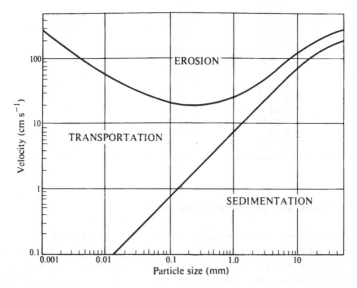

Figure 7.7 Hjulstrom (1935) relationship between particle size and velocities for erosion and sedimentation.

$$Re^* = du^*/v \,,$$

where d = particle size, u^* = shear velocity = $(\tau_0/\rho)^{1/2}$, ρ = density, v = kinematic viscosity.

For fully turbulent flow a value of $F_s \approx 0.06$ is a satisfactory approximation. The entrainment function can be used to estimate bedload transport from the empirical Einstein (1942) bedload function:

$$\Phi = q_s/wd \,,$$

where q_s = sediment discharge per unit width, d = grain size of sediment, Φ = the Einstein bedload function, w = particle fall velocity = $G(gd(S_s - 1))^{1/2}$ where

$$G = [0.6667 + \{36v^2/gd^3(S_s - 1)\}]^{1/2} - \{36v^2/gd^3(S_s - 1)\}^{1/2} \,;$$

$G \approx 2/3$ when $d \geq 1.6$ mm for grains of specific gravity 2.6 in water; g = the gravitational constant, S_s = the solid : fluid density ratio, v = kinematic viscosity (0.013 cm² s⁻¹ at 10 °C; 0.008 cm² s⁻¹ at 30 °C).

This function is plotted as a function of F_s in Figure 7.10, after Henderson (1966).

Estimates of bedload transport may also be obtained from the relationships of Bagnold (1956) and Meyer-Peter and Muller (1948). These are

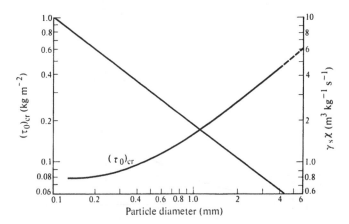

Figure 7.8 Evaluation of parameters in the Duboys equation. γ_s = unit weight of sediment.

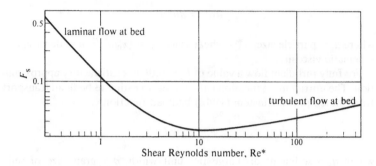

Figure 7.9 Shields' (1936) entrainment function, F_s (dimensionless critical shear stress), as a function of shear Reynolds number.

shown in Figure 7.11, after Carson and Kirkby (1972). Dimensionless shear, θ, is given by

$$\theta = Rs/\Delta d \ ,$$

where R = hydraulic radius, s = slope of the bed, $\Delta = (\rho - \rho_w)/\rho_w$, ρ = density of sediment, ρ_w = density of water, d = grain diameter.

The Bagnold function is proportional to a parameter B. The main curves shown are for $B = 1.0$ and the inset shows B as a function of grain diameter. c is the concentration of sediment. Dimensionless sediment

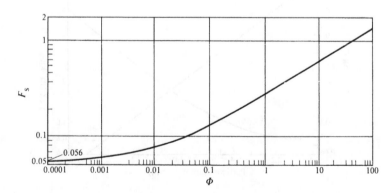

Figure 7.10 Einstein's (1942) bedload function, Φ, as a function of Shields' entrainment function, F_s (after Henderson 1966).

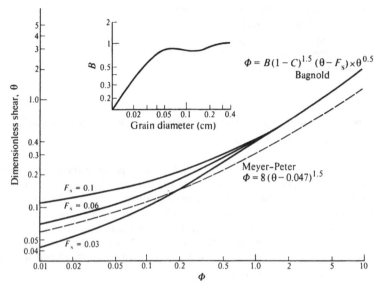

Figure 7.11 Bagnold (1956) and Meyer-Peter and Muller (1948) sediment transport functions. The inset is explained in the text.

transport (φ) is related to sediment transport per unit width (s) by

$$\varphi = s/(g \, \Delta d^3)^{1/2} \, .$$

A further, more complex, development of this approach is given by Bagnold (1980).

An analytical bedload function has also been given by Einstein (1950). This is probably the most generally applicable but is also the most involved and is not suitable for field evaluation. A modified Einstein procedure has been devised by which total load can be evaluated and, for measurements that must be determined in the field to allow computations by this method, a checklist is given below after Colby and Hubbell 1961 and Garde and Ranga Raju 1977:

(1) stream width;
(2) mean depth;
(3) mean velocity;
(4) size analysis of bed material;
(5) water temperature;

(6) for a sample of suspended sediment;
    (a) particle size analysis,
    (b) depth at the sampling vertical,
    (c) concentration of suspended sediment,
    (d) sampled depth.

7.6 Relationships between bedforms, flow characteristics and particle sizes

The material in this section is included to allow preliminary examination and interpretation of current processes and palaeohydraulic features to be made in the field. Many other relationships have been suggested, for which reference should be made to Allen (1965, 1969), Dury (1976), Jopling (1966), Southard (1971) and Etheridge and Schumm (1978).

Figure 7.12 (after Simons & Richardson 1971) shows the occurrence of bedforms in terms of the median fall diameter of bedload (d) and stream power (P), where P is given by

$$P = \gamma DSV \,,$$

where γ = unit weight of water (kg m^{-2}), D = depth of flow (m), S = channel slope (as gradient, e.g. mm^{-1}), V = velocity (m s^{-1}).

This can be used to predict the likely bedforms resulting from given flow conditions, or to estimate stream power and hence D, S or V from sedimentary structures. Alternative interpretations are given by Southard (1971), Allen (1969) and Bogardi (1961).

Bedforms and flow characteristics associated with increasing velocity of flow are shown in Figure 7.13a for the lower flow regime and in Figure 7.13b for the upper flow regime after Harms and Fahnestock (1965). These figures can be used to give crude estimates of flow resistance from other observed characteristics or to predict the bedforms likely to be produced by particular flow conditions.

In palaeohydraulic interpretation of sediments, maximum particle size may be used to estimate shear stress during motion from Figure 7.14, after Baker and Ritter (1975). Many limitations apply to the relationship shown in this figure since it represents a compilation from only a limited number of studies available to the original authors. In particular there is a considerable spread of values about the regression relationship indicated. Once shear stress has been determined from the figure, channel depth can be estimated from the relationship

$$\tau_0 = \gamma RS \, ,$$

where τ_0 = shear stress, γ = specific weight of fluid ≈ 1000 kg m^{-3} for water without appreciable sediment in suspension, R = hydraulic radius (m) $\approx$ depth for wide rivers, S = energy slope $\approx$ slope of the channel $\approx$ slope of the deposit.

7.7   Conductivity measurements in water quality analysis

This section includes information allowing field measurements of conductivity to be corrected for temperature and pH, and conductivity meters to be checked. For more detailed information see Edwards *et al.* (1975) and Finlayson (1979).

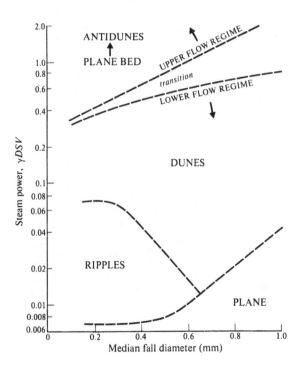

Figure 7.12   Relationship between median fall diameter of bedload and bedforms (Simons & Richardson 1971).

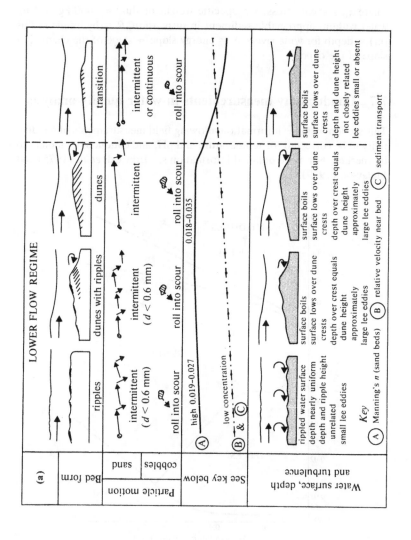

(a) LOWER FLOW REGIME

Bed form		ripples	dunes with ripples	dunes	transition
Particle motion	cobbles sand	intermittent ($d < 0.6$ mm) roll into scour	intermittent ($d < 0.6$ mm) roll into scour	intermittent roll into scour	intermittent or continuous roll into scour
See key below		high 0.019–0.027		0.018–0.035	
Water surface, depth and turbulence		rippled water surface depth nearly uniform depth and ripple height unrelated small lee eddies	surface boils surface lows over dune crests depth over crest equals dune height approximately large lee eddies	surface boils surface lows over dune crests depth over crest equals dune height approximately large lee eddies	surface boils surface lows over dune crests depth and dune height not closely related lee eddies small or absent

Key

(A) Manning's *n* (sand beds) (B) relative velocity near bed (C) sediment transport

(A) high · (B) & (C) low concentration

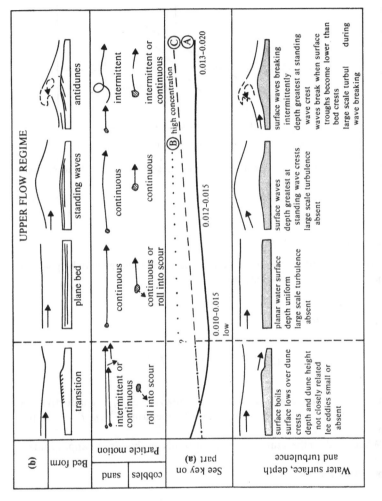

Figure 7.13 Flow characteristics, sediment motion and bedforms associated with increasing velocity of flow (Harms & Fahnstock 1965): (a) lower flow regime; (b) upper flow regime.

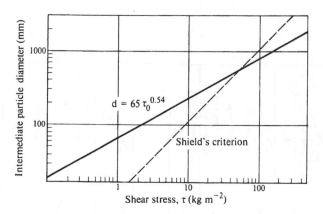

Figure 7.14 Particle size versus mean shear stress for coarse bedload materials (Baker & Ritter 1975).

Correction for temperature

If not performed automatically, temperature corrections may be given by one of the following two methods.

Approximate method

$$L_R = L_T - 0.02(T - R)L_T \, ,$$

where L_R = conductivity at reference temperature, L_T = conductivity at sampled temperature, R = reference temperature (°C), T = sample temperature (°C).

When field temperatures are near zero this method is not appropriate.

More exact method

$$L_R = L_T/\{1 + a(T - R)\} \, ,$$

where a = a temperature coefficient (in % °C^{-1}). a lies between 1.5 and 2.4% °C^{-1}, and may be determined experimentally by taking conductivity measurements at two temperatures, t_1 and t_2, and calculating a from

$$a = (L_{t_2} - L_{t_1})/(L_{t_1}(t_2 - R) - L_{t_2}(t_1 - R)).$$

For laboratory purposes 25 °C is used as the reference temperature; for field use 10 °C may be more appropriate.

Correction for pH

The hydrogen ion concentration influences the conductivity of water greatly and, at pH values lower than about 5.5 (i.e. in acidic waters), it is necessary to correct values for conductivity due to hydrogen. The amount of conductivity to be subtracted at a reference temperature of 25 °C is shown in Figure 7.15 (after Finlayson 1979) for measured values of pH.

Calibration and checking of conductivity meters

Readings of conductivity meters may be checked in the field against known solutions. The table below gives specific conductances for varying concentrations of KCl at 25 °C; 0.1 mol KCl = 7.455 g l^{-1}.

Concentration (mol)	Specific conductances ($\mu mho\ cm^{-1}$)
0.0001	14.94
0.0005	73.90
0.001	147.00
0.005	717.80
0.01	1413.00
0.1	12 900.00
1.0	111 900.00

Table 7.6 Constants for use in hydraulic calculations.

Water temperature (°C)	Specific gravity	Specific weight ($N\ m^{-3}$)	Viscosity ($N\ s\ m^{-2}$)	Kinematic viscosity ($m^2\ s^{-1}$)
0.0	0.9999	9805.4	0.001 794	0.000 001 794
4.0	1.0000	9806.5	0.001 568	0.000 001 567
10.0	0.9997	9803.8	0.001 310	0.000 001 310
15.6	0.9990	9797.5	0.001 129	0.000 001 131
21.1	0.9980	9786.6	0.000 982	0.000 000 984
26.7	0.9966	9774.0	0.000 861	0.000 000 864
37.8	0.9931	9739.4	0.000 682	0.000 000 687

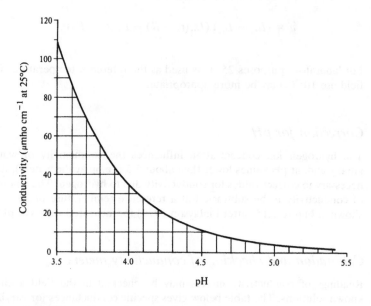

Figure 7.15 Correction to be applied to conductivity measurements for pH (Finlayson 1979).

8 Glacial processes

This chapter contains a range of advice on simple techniques that may be used by the geomorphologist in areas of active glaciation, as reviewed in Goudie (1981), pp 213–25. The first three sections deal essentially with survey problems. These include provision of fixed points (Sec. 8.1) and stake networks (Sec. 8.3) for monitoring glacier mass balance, surface velocity and surface strain rates. Section 8.2 deals with mapping snow and ice. Section 8.4 gives details of a simple method of establishing the surface velocity of the glacier, and means of establishing the orientation of the three principal axes of strain and the magnitude of the strain rates along them is covered in Section 8.5. The measurement of the accumulation on the glacier is covered in Section 8.6. The rammsonde may be used to investigate the variation of hardness of snow with depth and, under favourable conditions, to locate the previous summer surface without the need to dig snow pits. When calibrated against density it may be used to investigate the variation of density in the firn area. The ram hardness and its relation to strength are covered in Section 8.7. Ablation measurements are covered in Section 8.8. The minimum meteorological observations that should be made in any glacier study are briefly listed in Section 8.9 and a few notes on the minimum observations of water discharge are given in Section 8.10. Reference should be made to Chapter 7 for fuller details of stream gauging and sediment-yield measurement procedures. Chapter 12 provides some brief notes on the requirements for water samples for chemical or physical analysis. Measurements of the temperature of glacier ice are important to most glaciological studies but ideally require the use of pre-calibrated thermistor probes which can be frozen into a borehole. Papers by Harrison (1975), Classen (1977) and Hooke and Koci (1978) give examples of this method. A crude measurement of temperatures in an open borehole in a glacier may be made by lowering a thermometer or temperature probe down the hole and discouraging convection by placing discs on the wire both above and below the probe (Thomas 1976).

8.1 Choosing and marking fixed points for glacier surveys

Østrem and Stanley (1966) recommend that key points on the glacier surface should be marked so that they are easily visible from survey stations on stable ground overlooking the glacier and so that they show up on photography. Similarly the fixed points overlooking the glacier should be easily picked out both from the glacier and on any aerial photographs that may be taken.

Fixed points surrounding the glacier should be chosen so that they are readily visible both from the glacier and from other fixed points. If possible, prominent topographic features should be chosen, although ease of access should be a primary consideration. If the chosen point is not easily picked out, a cairn surmounted by a flag should be constructed. So that the chosen point shows up on aerial photographs or when viewed from above it should be marked with pieces of white cloth at least 1 m in width and 5 m long, placed in the form of an 'L' with the inner corner of the 'L' at the fixed point.

Key points on the glacier need to be marked differently depending on whether they are in the accumulation or ablation area. In the ablation area large flags, white on a dirty glacier and black on a clean glacier, may be attached to the stakes used for various measurements. To help recognition of stakes when viewed from above, their locations may be marked by placing two rows of boulders on the glacier surface with the intersection of them at the stake. The orientation of these rows should be chosen so that it does not coincide with any natural glacier surface features. In the accumulation area powdered dye or lampblack may be used to mark a circle or semicircle c. 10 m in radius and 1 m wide around the chosen fixed point.

As a means of supplementary control, distances between fixed points should be measured on the ground so that, in the event of markers becoming obscured, the positions of the fixed points can be relocated.

8.2   Mapping snow and ice

The symbols given in Figure 8.1 may be used to map the boundaries between different ice types. Symbols for morainic features are given in Chapter 2. Surface snow may be fully described using the scheme given in Section 8.6 and Unesco/IASH/WMO (1970) gives a comprehensive set of symbols that permit these data to be both mapped and displayed as profiles. Brown should be used for drift and bedrock features and blue for depicting ice and snow features on a map.

8.3   Recommendations for placing stakes on a glacier surface

Stakes inserted into the surface of a glacier may be used for a number of

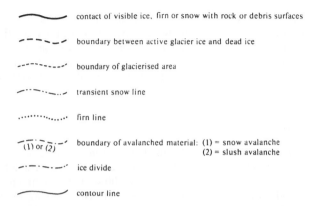

contact of visible ice, firn or snow with rock or debris surfaces

boundary between active glacier ice and dead ice

boundary of glacierised area

transient snow line

firn line

boundary of avalanched material: (1) = snow avalanche
(2) = slush avalanche

ice divide

contour line

Figure 8.1 Symbols for mapping ice features (after Blachut & Muller 1966).

purposes, among which are the measurement of accumulation, ablation, glacier surface velocity and surface strain rates. In the first two cases the vertical stability of the stakes is critical and in the last two cases care must be taken to ensure perfect coupling with the moving glacier ice. The recommendations that follow are based on those of Østrem and Stanley (1966).

Cover of the network

Where stakes are to be used for the determination of the mass balance of the glacier, there should ideally be an even and fairly dense cover over the whole of the glacier surface. This is rarely a practical proposition and, as an alternative, stakes may be placed in a regular geometric pattern. On valley glaciers it is convenient to place a long line of stakes up the centre of the glacier with transverse lines at regular intervals. These transverse lines should be at right angles to the centre line. The accuracy of the results will depend to a very large extent on the efficiency of the sampling network. The number of sampling points will be largely determined by the time taken to emplace and read the stakes and the other work that is to be undertaken. In general, a greater density of stakes should be used in the accumulation area since variations in both accumulation and ablation are greater there than elsewhere. On most glaciers the pattern of accumulation and ablation shows some consistency from year to year and in a continuing study a reduction in the number of sampling points may be made to those that appear to be representative of wide areas. The cover of stakes required for surface velocity measurements will normally consist of a series of transverse lines distributed evenly in both the accumulation and ablation zones. Stakes for the determination of the orientation and magnitude of the three principal axes of strain should be laid out as described in Section 8.5 and the centre stake may be used to measure surface velocity.

Numbering

The centre line stakes should be numbered from the snout towards the firn area, using the ascending numbers 100, 200, 300, etc. The left-hand limb of the transverse line centred on stake 200 should be numbered using ascending odd numbers away from stake 200. Thus moving to the left from stake 200 successive stakes are numbered 201, 203, 205, etc. The right-hand limb should be similarly numbered using even numbers. Additional longitudinal and transverse stakes are readily inserted into this scheme and the logical layout makes it possible to find poorly exposed stakes.

Replacement of missing stakes

New stakes should be inserted as close as possible to the site of the missing stake and numbered with the original number plus a prefix indicating the generation of replacement. For example, the first replacement for stake 103 would be 1103, the second generation 2103, etc. Replacement stakes should be removed from the glacier as soon as the old stake is rediscovered.

Duplication of stakes

When a stake is difficult to relocate, a duplicate may be placed nearby at a known distance from the original to aid location. The duplicate should be well flagged and labelled with the original number plus a prefix. Thus A103 points the way to 103. A second duplicate would be B103, etc.

Stake extension

Tubular aluminium stakes may be extended by inserting a 30 cm long steel pipe into the top and adding a 2 m extension tube. The extension should be numbered. Thus the first extension to stake 103 becomes 103/1, the second 103/2, etc.

Inserting stakes

Aluminium is the best general material for snow stakes although steel stakes may be required in humid climates where strong winds and the weight of hoar frost make stronger materials necessary. The optimum length is 4 m or 5 m. Stakes should be cut to an exact length and marked right around the circumference at every metre to facilitate reading. Ideally the marking should be a shallow turned groove.

In the firn area the stakes should be pushed into the snow until they rest on the previous summer surface. Vertical settling may be prevented or at least minimised by placing a cork or wooden bung in the foot of the stake before pushing it home. If a large-diameter hole is drilled, a disc of diameter greater than the stake may be placed in the hole before inserting the stake.

When stakes are inserted in the glacier ice a hole must be drilled and the stake allowed to freeze into place. Sticking of the ice auger may be prevented by pouring antifreeze down the hole. It is best to drill when cold as this prevents melt water percolating down the hole and freezing.

When stakes are inserted at the beginning of the ablation season, only 10–20 cm need be left showing. If placed near the end of the ablation season, 2–2.5 m should be left exposed. When less than 1.5 m of the stake remains frozen solidly in place the stake should be replaced. Replacement stakes should be placed 1 m up-glacier from the original. If the stakes are used for glacier movement studies, the precise location of the new stake must be recorded well before the old one comes loose.

8.4 Measurement of glacier surface velocity

Measurement of the surface velocity of a glacier may be accomplished using standard surveying techniques to measure the horizontal and vertical components of velocity (Paterson 1969). Markers, usually wooden or aluminium stakes, are set in the ice and their position surveyed in at fixed time intervals. Stakes should be set out in the pattern recommended for mass balance measurements. In the ablation area care must be taken to drill holes deep enough to set the stakes so that they do not melt out during the summer. In the accumulation area care must be taken to anchor the stakes firmly in position to enable measurements of the vertical component to be made. Velocity measurements may be conveniently related to strain rate determinations (Sec. 8.5) by using the centre stake of the square array for the velocity measurements.

Figure 8.2 shows the measurements required. The horizontal and vertical displacements (ΔX and ΔY respectively) of the top of the stake are measured with respect to axes fixed in space. The horizontal component of velocity (U) is equal to ΔX. This can be monitored using repeated triangulation over periods of a few days. In the ablation area the ice flow is upwards and the vertical displacement is given by

$$\Delta Y = \Delta X \tan \alpha - V ,$$

where V is the amount of upward movement of the glacier surface per unit time. Thus the vertical component of flow (V) is

$$V = \Delta X \tan \alpha - \Delta Y .$$

The vertical component is small and successive elevation measurements of the marker stake should be separated by at least several months. The component perpendicular to the surface (v) is given by

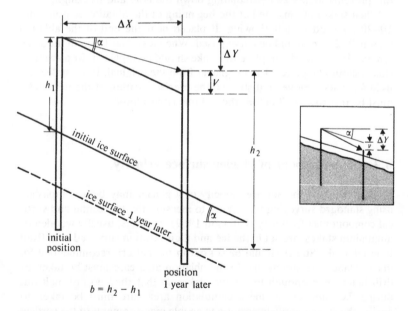

$b = h_2 - h_1$

Figure 8.2 Measurements required for the determination of glacier velocity. This figure shows the situation for stakes placed in the ablation area. The inset shows the situation in the accumulation area and the measurements and calculations should be adjusted accordingly (after Paterson 1969).

$$v = -V \cos \alpha$$
$$= \Delta Y \cos \alpha - \Delta X \sin \alpha$$

(v is normally positive downwards). At a point off the centre line of the glacier X is equal to the component of horizontal velocity in the direction of the centre line (U).

$$U = \Delta X \sec A$$

where A is the angle between the direction of the horizontal velocity and the centre line. The ice surface at the stake would have an elevation change over a unit time of Δh: $\Delta h = b - v$ where b is the mass balance as ice thickness for the unit time.

8.5   Determination of the strain rate at the surface of a glacier

Nye (1959) gives details of the measurements and calculations required to determine the magnitude and orientation of the three principal strain axes at the surface of a glacier and the strain rates along them. A square pattern consisting of five stakes should be laid out on the glacier surface as shown in Figure 8.3. The x axis of the array is along one of the diagonals and points down the glacier. The z axis is transverse to the glacier flow direction and should be at right angles to the x axis. The y axis is normal to

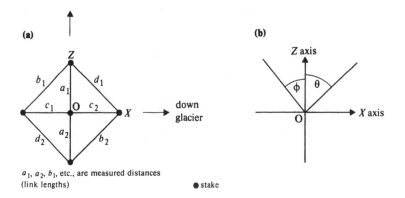

a_1, a_2, b_1, etc., are measured distances
(link lengths)   ● stake

Figure 8.3   Stake layout and notation for strain determination (after Nye 1959).

the glacier surface. The stake pattern should be as near as possible to a perfect square and the orientation as well as lengths of all the links should be recorded.

The position of the top of the centre stake should be fixed in both horizontal and vertical planes by theodolite triangulation and/or resection to fixed points on the glacier margin. This allows for the calculation of the rate of movement at the strain rate determination site by the method described in Section 8.4.

The distance between the stakes, the link lengths, and the orientation of each of the links should be monitored at suitable intervals. The time interval between measurements will vary depending on the scope of the investigation, the time available for that task, the means of measurement available and the resolution required of temporal variations in the rates of strain. Nye (1959), working on Austerdalsbreen in Norway, made repeated measurements at intervals of c. 5–6 days. Waag and Echelmeyer (1979), working in Alaska, used an interval of 11 days. Very small strain rates may require intervals of up to a year between measurements, although with sophisticated measurement methods accurate determinations may be made over 1 day (Holdsworth 1975). Link lengths should for most purposes be at least 50 m and ideally at least 100 m. Measurements should be made using a steel tape, taking care to keep the tension constant for each measurement. Measurements should be made to precisely the same point on the stake on each occasion. The heights of the stakes and their tilt should also be recorded. If stakes need to be rebored into the ice, they should be resurveyed both immediately before and after reboring. The stakes must be held firmly in the ice and should be rebored well before they come loose.

From the measurement of the eight link lengths ($a_1, a_2, b_1, b_2, c_1, c_2, d_1$ and d_2) eight values of the strain rate ($\dot{\varepsilon}$) may be obtained from the equation

$$\dot{\varepsilon} = (1/\Delta t)\ln(l_2/l_1),$$

where l_1 and l_2 are the initial and final link lengths for each of the links in turn and Δt is the time interval between measurements.

The eight values of $\dot{\varepsilon}$ may be reduced to four values of $\dot{\varepsilon}$ corresponding to the directions $\theta = 0°, 45°, 90°$ and $135°$, where θ is measured clockwise from OZ (Fig. 8.3b):

$$\dot{\varepsilon}_0 = (a_1 + a_2)/2,$$
$$\dot{\varepsilon}_{45} = (b_1 + b_2)/2,$$

$$\dot{\varepsilon}_{90} = (c_1 + c_2)/2,$$
$$\dot{\varepsilon}_{135} = (d_1 + d_2)/2.$$

These four values refer to the point O at the centre of the square based on the assumption that the strain rate varies linearly over the area of the square.

The residual (v) should be calculated from

$$v = (\dot{\varepsilon}_0 + \dot{\varepsilon}_{90} - \dot{\varepsilon}_{45} - \dot{\varepsilon}_{135})/4.$$

Then

$$\dot{\varepsilon}_x = -1/4\,\dot{\varepsilon}_0 + 1/4\,\dot{\varepsilon}_{45} + 3/4\,\dot{\varepsilon}_{90} + 1/4\,\dot{\varepsilon}_{135},$$
$$\dot{\varepsilon}_{zx} = \qquad\qquad 1/2\,\dot{\varepsilon}_{45} \qquad\quad - 1/2\,\dot{\varepsilon}_{135},$$
$$\dot{\varepsilon}_z = 3/4\,\dot{\varepsilon}_0 + 1/4\,\dot{\varepsilon}_{45} - 1/4\,\dot{\varepsilon}_{90} + 1/4\,\dot{\varepsilon}_{135}.$$

The standard errors (S.E.) are as follows:

$$3^{0.5}|v| \qquad \text{for } \dot{\varepsilon}_x;$$
$$3^{0.5}|v| \qquad \text{for } \dot{\varepsilon}_z;$$
$$2^{0.5}|v| \qquad \text{for } \dot{\varepsilon}_{zx}.$$

The magnitudes and directions of the principal strain rates ($\dot{\varepsilon}_1$, $\dot{\varepsilon}_2$ and $\dot{\varepsilon}_3$) can be calculated from

$$\dot{\varepsilon}_1 = 0.5(\dot{\varepsilon}_x + \dot{\varepsilon}_z) - \{0.25(\dot{\varepsilon}_x - \dot{\varepsilon}_z)^2 + \dot{\varepsilon}_{zx}^2\}^{1/2},$$
$$\dot{\varepsilon}_2 = -(\dot{\varepsilon}_x + \dot{\varepsilon}_z),$$
$$\dot{\varepsilon}_3 = 0.5(\dot{\varepsilon}_x + \dot{\varepsilon}_z) + \{0.25(\dot{\varepsilon}_x - \dot{\varepsilon}_z)^2 + \dot{\varepsilon}_{zx}^2\}^{1/2}$$

The principal strain rate $\dot{\varepsilon}_2$ lies in the vertical plane and is $\dot{\varepsilon}_y$. The other two principal strain rates $\dot{\varepsilon}_1$ and $\dot{\varepsilon}_3$ lie in the horizontal X, Z plane.

The principal strain rate axis is that of $\dot{\varepsilon}_3$ if $\dot{\varepsilon}_z > \dot{\varepsilon}_x$ or $\dot{\varepsilon}_1$ if $\dot{\varepsilon}_z < \dot{\varepsilon}_x$.

The angle between OZ and the principal axis (φ) is given by

$$\tan 2\varphi = 2\dot{\varepsilon}_{zx}/(\dot{\varepsilon}_x - \dot{\varepsilon}_z), \qquad -\pi/4 < \varphi < \pi/4.$$

The sense of φ is defined in Figure 8.3.

8.6 Accumulation measurements

Accumulation on a glacier may be measured either by snow depth soundings or from snow pit studies.

Snow depth soundings

In order to assess adequately variable snow depths, a sampling density of
about 100 points per kilometre should be aimed at. Ideally the points
should be evenly spread over the glacier or placed in a series of sounding
profiles. It is convenient to place the profiles between snow stakes laid out
as described in Section 8.3. Along the profiles soundings should be made
at intervals of 50 m. Soundings should be made to the nearest centimetre.
Measurements of snow depth should be plotted up immediately in order
to ascertain the need for additional soundings in areas where snow depth
is variable.

It is best to start the soundings at the snout since the snow cover is
thinnest and the previous summer surface will be glacier ice and therefore
well defined. Higher up the glacier the summer surface may be poorly
defined, especially after a cool summer, and it may be advantageous to
mark the current year's summer surface to facilitate the following year's
survey. Soundings in the accumulation zone take longer to carry out and
are quantitatively most significant. Up to three-quarters of the total survey time should be spent in this area.

Snow pit studies

Full details of snow pit studies are beyond the scope of this work. The
following are basic guidelines only. The user of these methods should be
conversant with the main processes by which snow is transformed into
glacier ice. Brief outlines of these are given by Seligman (1936) and
Paterson (1969); Armstrong *et al.* (1966) provide a useful illustrated
glossary of terms for types of snow and ice.

The number of pits dug will depend on the purpose of the work. A bare
minimum of three should be dug, one low in the firn area, one in the
middle and one high up. Before starting digging use a snow probe to
determine snow thickness. Continue digging to *c*. 0.5 m below the previous summer surface. Decide before digging which face of the pit is to
remain untouched for sampling and measurements. This should preferably be the most shaded face. The pit should be sufficiently large to give a
working space at least 1 m² at the bottom. If the pits are to be dug near
existing stakes they should all be at a consistent distance downstream from
the stakes. Five or 10 m is recommended. If there is no stake at the pit
location, one should be inserted to aid relocation should it be necessary.
As soon as possible after digging, snow temperature readings should be
made to ascertain whether melting has occurred. Snow density at levels
down to 0.5 m below the previous summer surface should be determined

using a standard snow sampler and balance. Typical densities for snow and ice are given in Table 8.1. Table 8.2 gives a checklist of features of the snow surface that should be recorded. For layers of deposited snow the density, free water content (Table 8.3), the grain size and shape (Tables 8.4 & 5), the strength (Table 8.6), the ram hardness (Table 8.7), the snow temperature and the presence of impurities or ice layers should be noted.

The previous summer surface may be marked by a layer of dirt, or by a sudden change in density, hardness or grain shape. In the case of arctic glaciers the base of a layer of depth hoar may be taken as the summer surface (Paterson 1969).

Table 8.1 Typical snow and ice densities (Paterson 1969).

Snow/ice type	Density (g cm^{-3})	Snow/ice type	Density (g cm^{-3})
new snow	0.01–0.07	wind-packed snow	0.35–0.4
damp new snow	0.01–0.2	firn	0.4–0.85
settled snow	0.2–0.3	very wet snow and firn	0.7–0.8
depth hoar	0.2–0.3	glacier ice	0.85–0.91

Table 8.2 Checklist for the description of snow surface conditions.

Nature of the surface deposit

surface hoar	a surface layer of ice crystals mainly in the form of scales, needles and feathers formed by condensation on the snow
rime	a white fluffy layer of relatively sparse leaflets and tufts of plate crystals formed by sublimation; state whether hard or soft
glazed frost	an apparently structureless, turbid or glassy amorphous-looking ice formed by rapid freezing of supercooled water

Surface roughness
State whether smooth, wavy, concave furrows, convex furrows, or random furrows
Surface penetrability
Should be measured by noting the depth of imprint in the snow surface formed by the weight of one person supported on one foot or on one ski

Table 8.3 Assessment of the free water content of snow (Unesco/IASH/WMO 1970).

Category		Characteristics
dry	(Wa)	snowball cannot be made
moist	(Wb)	liquid water not obvious but snowball can be made
wet	(Wc)	liquid water is visible
very wet	(Wd)	liquid water abundant but does not drain freely
slush	(We)	liquid water drains out of snow with slight pressure

Table 8.4   Assessment of the grain size of snow (Unesco/IASH/WMO 1970).

Category		Size range (mm)
very fine	(Da)	<0.5
fine	(Db)	0.5–1.0
medium	(Dc)	1.0–2.0
coarse	(Dd)	2.0–4.0
very coarse	(De)	>4.0

Table 8.5   Assessment of the grain shape of snow (Unesco/IASH/WMO 1970).

Category		Characteristics
new snow	(Fa)	unchanged or slightly transformed snow crystals; original shapes well recognisable
often felt-like	(Fb)	crystals in advanced transformation but elements of original crystals still recognisable
granular, rounded	(Fc)	rounded, often elongated grains; with or without melting
granular with facets, full crystals	(Fd)	often rounded grains or cup-shaped elements intermixed
depth hoar	(Fe)	cup-shaped crystals, but not necessarily fully developed; usually only fragments of cups, characterised by re-entrant angles and peculiar ledges

Table 8.6   Assessment of snow strength (Unesco/IASH/WMO 1970).

Category		Characteristics
very soft	(Ka)	can be penetrated by gloved fist
soft	(Kb)	can be penetrated by the tips of four fingers of a leather-gloved hand
medium hard	(Kc)	can be penetrated by one finger of a leather-gloved hand
hard	(Kd)	can be penetrated by a pencil
very hard	(Ke)	can be penetrated by a knife

8.7   Ram hardness of snow

This measurement of snow hardness and strength is based on the amount of penetration of a standard cone-penetrometer. The cone should have an apex angle of 60° and a diameter of 40 mm.

$$\text{Ram hardness, } R = (w + qQ + P) + (Phn/x) \text{ kg force .}$$

Table 8.7 Ram hardness, shear strength and descriptive terms (Unesco/IASH/WMO 1970).

Category		Ram hardness, R(kgf)	Shear strength, $K(g\,cm^{-2})$
very soft	(Ra)	0–2.0	0–10.0
soft	(Rb)	2.0–15.0	10.0–75.0
medium	(Rc)	15.0–50.0	75.0–250.0
hard	(Rd)	50.0–100.0	250.0–500.0
very hard	(Re)	100.0	500.0

Where w = weight of the first element of the penetrometer (kg), q = number of 1 m extension rods, Q = weight of a 1 m extension rod, P = weight of the hammer, h = height of the hammer drop (cm), n = number of times the hammer is dropped per unit depth, x = penetration of the rod assemblage (cm). Interpretation is facilitated if the results are plotted as a cumulative product of the depth interval and the ram hardness value; i.e. $\Sigma \Delta x R$ against depth. A series of verbal terms and the corresponding ram hardness values are given in Table 8.7. Also given in this table are approximate values of shear strength.

Gubler (1975) has proposed a modified formula for the ram hardness that more realistically takes account of energy transfer to the rammsonde point. An empirically derived correction factor (C) depending on the ratio of hammer fall to penetration depth must be applied to the last term of the equation given above. Thus

$$R_c = (w + qQ + P) + C(Phn/x) \text{ kg force} .$$

The value of C, the correction factor, may be read from the graph in Figure 8.4.

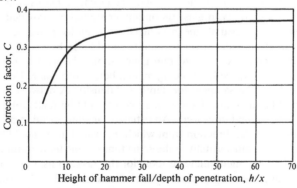

Figure 8.4 Correction factors for ram hardness equation (after Gubler 1975).

8.8 Ablation measurements

Most ablation studies are carried out by repeated observation of surface lowering at stakes set into the glacier surface. In order to obtain consistent results the following rules should be observed.

Where the stakes are drilled into ice, the measurement should be taken from the top of the stake down to the ice surface. The ice surface should be defined by the lower edge of an ice axe placed on the ice transverse to the flow direction and touching the stake.

Where snow still covers the ice the measurement must include the visible length of the stake and the snow depth. Snow depth should be taken as an average of three values of depth determined by probing within 1–2 m of the stake. The height of the snow surface should be defined as in the paragraph above. Surface snow should be sampled regularly during the summer for density determination.

8.9 Meteorological observations

Observations should be made early in the morning and in the evening so as not to hinder glaciological work. It is desirable for the times of observations to be as close to the local standard as possible. The following minimum observations should be made:

(a) *Air temperature and humidity* Ideally a thermohygrograph in a Stevenson screen should be used. Time check marks and a check against a mercury in glass thermometer should be provided.
(b) *Cloud cover* A daily mean cloud cover in tenths should be recorded. Cloud altitude and cover variability should also be noted.
(c) *Precipitation* A standard rain gauge should be used to record rain and wet snow, the latter being melted before recording. Heavy snow falls should be sampled for depth and density. A network of gauges should be maintained and at least one should be read daily.
(d) *Wind speed and direction* An estimate of wind speed in metres per second and the direction from which the wind blows should be made twice a day, most usually at the same times as the temperature is read. A hand-held anemometer is ideal for this task. In its absence a small piece of paper may be released and allowed to fall to the ground and its speed roughly calculated.

(e) *Net radiation*   Ideally a net radiometer should be used, but in its absence sunshine hours should be recorded and some estimation made of the glacier surface reflection coefficient. Empirical relations may then be used to determine incoming radiation (Black *et al.* 1954).

8.10   Water discharge measurements

Whenever possible water and sediment discharge from a glaciated basin should be monitored. Discharge is usually recorded daily at the same time as meteorological observations and is most conveniently accomplished by using a rated section. Methods of discharge measurement for rating a cross section are given in Chapter 7. Suspended sediment load and bed-load should also be monitored on a daily basis and is also covered in Chapter 7.

Additional readings of discharge should be made in the late morning and early evening wherever possible in order to characterise the normal diurnal variation in glacial streams. Load observations should also be made at these times if possible. It is highly desirable to monitor at least one major event on an hourly basis if at all possible. Similarly, hourly readings throughout a 'normal' 24 h period should be made to enable water and sediment discharge curves to be drawn based on three or four readings per day when conditions are 'normal'.

9 Aeolian processes

This chapter contains material appropriate to the examination and analysis of wind processes, as reviewed in Goudie (1981), pp. 226–46. Basic relationships governing entrainment and transport by the wind are given in the first section as a basis for field estimates and analysis. The second section gives an outline of the characteristics of aeolian sediments to assist in the identification of deposits and material allowing classification of aeolian depositional landforms.

9.1 Relationships concerning wind velocity, drag velocity, sediment entrainment and transport

The relationships given in this section are included as a basis for preliminary examination and analysis of field data. More detailed evaluation of these relationships may be found in texts such as those of Bagnold (1954) and Cooke and Warren (1973). This section also includes a graph relating visibility to dust concentration in the atmosphere which may be used for field estimation of dust concentrations.

Drag velocity (V_*) or velocity gradient is given by

$$V_* = KV_z/\log(z/k) ,$$

where V_z = velocity at height z, k = height at which velocity is zero, K = von Kàrmàn constant, ≈ 0.04, but which varies with the temperature gradient. V_z may be derived from

$$V_z = 5.75\, V_* \log(z/k) ,$$

where V_* = the drag velocity, k = surface roughness $\approx 0.081(\log d)/0.18$ for sandy surfaces, (where d = the mean particle diameter, in centimetres), $\approx d/30$ for relatively smooth surfaces, and tends towards $d/10$ for rough surfaces.

V_{*t}, the critical velocity at which particle motion begins (Bagnold 1954), or threshold drag velocity, may be derived from

$$V_{*t} = A\{(\sigma - \rho)gd/\rho\}^{1/2} ,$$

where σ = specific gravity of grains (g cm^{-3}), ρ = specific gravity of fluid (air) $\approx 1.22 \times 10^{-3}$ g cm^{-3}, g = gravitational constant (cm s^{-2}), d = mean particle diameter (cm). Bagnold's A is a coefficient which for grains above 0.1 mm in diameter ≈ 0.1. It is a function of Reynolds number (B), which is given by

$$B = V_{*t} r / v ,$$

where r = surface roughness ($\approx d$), v = kinematic viscosity of air (cm^2 s^{-1}) = viscosity ÷ density. Figure 9.1 depicts the relationship between A and B, based upon experimental results of Greeley $et\ al$. (1974).

By combination of formulae for V_z and V_{*t},

$$V_z \approx 5.75\ A\{gd(\sigma -\iota \rho)/\rho\}^{0.5}\{\log(z/k)\} .$$

The impact threshold velocity (V_{*i}) below which motion ceases is given by

$$V_{*i} = 680d^{0.5}\log(30/d) .$$

The relationships between fluid threshold velocity, impact threshold velocity and particle size are shown in Figure 9.2, after Bagnold (1954).

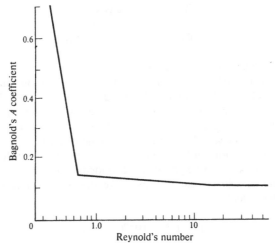

Figure 9.1 Relationship between Bagnold's A parameter and Reynolds number, based upon experimental results (Greeley $et\ al$. 1974).

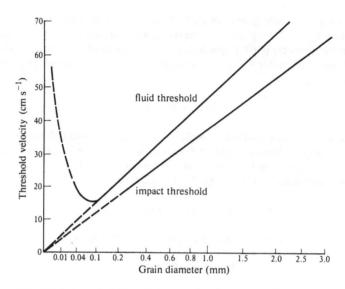

Figure 9.2 Variation of fluid and impact threshold velocities with grain size (Bagnold 1954).

Bagnold (1941) gives the following formula for estimation of total sand flow per unit time per unit of length transverse to motion:

$$q = C(d/D)^{0.5}(\rho/g)\alpha(V - V_t)^3,$$

where q = total load carried (tonnes per metre width per hour); C = a coefficient which takes the values, 1.5 for nearly uniform sand, 1.8 for naturally graded sand, 2.8 for poorly sorted sand, 3.5 for a pebbly surface; d = grain size (mm); D = standard grain size = 0.25 mm; ρ = specific gravity of air; g = gravitational constant; α = a constant = $0.174/\{\log (z/k)\}$; V = wind velocity at the measurement height; V_t = the fluid threshold velocity for sand movement; z = the height at which velocity is measured (m); k = the height at which velocity is zero.

For wind velocity measured 1 m above the bed the relationship for average dune sand is

$$q = 1.5 \times 10^{-9}(V - V_t)^3.$$

This relationship is depicted in Figure 9.3, and other suggested relationships are reviewed by Williams (1964) and Freyberger and Dean (1979).

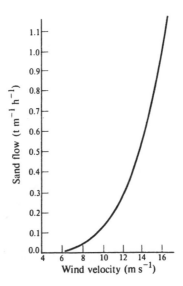

Figure 9.3 Relationship between sand flow rate and wind velocity measured at a height of 1 m above ground level, for 'average' dune sand (Bagnold 1954). This has a diameter of 0.25 mm and is normally graded. The quantity ρ/g for 'average' sand has a value of 1.25×10^{-6}.

Two versions of the relationship between visibility and dust concentration in the atmosphere are shown in Figure 9.4, after Chepil and Woodruff (1957) and Robinson (1968). These may be used to obtain estimates of dust concentrations from observed visibilities.

9.2 Aeolian deposits and depositional landforms

This section gives material of assistance in the examination of aeolian deposits and landforms. The characteristics of aeolian sediments are given below, in the form of a list (after McKee 1979, Glennie 1970) as an aid to the identification of deposits of possible aeolian origin. This is followed by material concerning depositional landforms.

Characteristics of aeolian sediments

(a) Sand bodies are usually light coloured and even-grained, with abrupt boundaries; often orthoquartzitic.

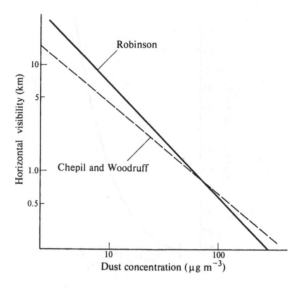

Figure 9.4 Relationship between dust concentrations at 1.8 m above ground level and horizontal visibility.

(b) Sequence of sands or sandstones varying in thickness from a few centimetres to several hundreds of metres.

(c) Foresets dip at angles from horizontal to 34° (less if compacted). Orientation of dips may be multi-modal or unimodal (see McKee 1966, 1979, Glennie 1970).

(d) Usually highly cross stratified, with tabular planar and wedge planar cross stratification being common and trough cross stratification occurring rarely (Fig. 6.11, Ch. 6).

(e) Laminae commonly planar, but ripples occasionally on steeply dipping foresets. Ripples invariably have high ripple indices (pp. 119–23, usually greater than 15) with parallel crests and troughs. According to Stone and Summers (1972) the ripple wavelength (L) is given by

$$L = 63.8 \, D^{-0.75} \, \text{cm} ,$$

where D = the mean grain diameter (mm).

(f) Individual laminae are well sorted, especially in the finer grain sizes. Sharp size differences commonly exist between maximum grain sizes of adjacent laminae.

(g) Grain size is commonly silt (60 μm) to coarse sand (2000 μm), with the bulk being 125–300 μm. Maximum grain size for aeolian transport is of the order of 1 cm, but grains over 5 mm are rare.

(h) Percentage of silt–clay generally well below 5% and even absent (other than authogenic clay).

(i) Larger sand grains tend to be well rounded.

(j) Clay drapes are very rare and are usually accompanied by evidence of water deposition.

(k) Quartz sands at shallow depths are commonly friable or lightly cemented with haematite. Local discoloration of red haematite-coated grains to green or white is not uncommon.

(l) Mica is generally absent.

(m) Adhesion ripples occur with an associated increase in clay content and the common presence of gypsum or anhydrite cement.

(n) Quartz grains which have not been cemented by calcite usually exhibit a frosted or pitted surface under the microscope.

Characteristics of aeolian stratification are described in Section 6.6, and more fully by Hunter (1977); McKee (1966, 1979) and Glennie (1970) consider the sedimentary characteristics of common dune types.

Classification of aeolian depositional features

Aeolian depositional features may be classified according to scale, as in Table 9.1. This classification contains an amalgam of data reported in Wilson (1970, 1972) and Cooke and Warren (1973), and probably represents the extreme size ranges of features likely to be encountered normally. Features of order two may be divided into free dunes and those related to obstacles. Table 9.2 and Figures 9.5 and 9.6 are concerned with

Table 9.1 Classification of aeolian depositional features according to scale.

Order	Name	Orientation	Wavelength	Amplitude
1	draa	longitudinal or transverse	300–5500 m	20–450 m
2†	dune	longitudinal or transverse	3–600 m	0.1–100 m
3	aerodynamic ripple	longitudinal or transverse	15–250 cm	0.2–5 cm
4	impact ripple	transverse	0.5–2000 cm	0.001–100 cm
	secondary ripple sinuosity	longitudinal	1–3000 cm	0.05–100 cm

† Features of order 2 are further subdivided in Table 9.2 and Figures 9.5 and 9.6.

Table 9.2 Terminology of basic dune types and other aeolian deposits (compiled from McKee 1979).

Form	Number of slipfaces	Name used in ground study of form, slipface and internal structure	Block diagram (see facing page)	Name used in space imagery and air-photo study of pattern and morphology
sheetlike with broad, flat surface	none	sheet	none, flat surface	sheet[†]
thin, elongated strip	none	stringer	none, flat surface	streak[†]
circular or elliptical mound	none[‡]	dome	(a)	dome-shaped
crescent in plan view	1	barchan	(b)	barchan
row of connected crescents in plan view	1	barchanoid ridge	(c)	barchanoid ridge
asymmetrical ridge	1	transverse ridge	(d)	not recognised
circular rim of depression	1 or more	blowout§	(e)	
'U' shape in plan view	1 or more	parabolic§	(f)	parabolic
symmetrical ridge	2	linear (seif)	(g)	linear
asymmetrical ridge	2	reversing	(h)	reversing
central peak with three or more arms	3 or more	star	(i)	star

crescentic (bracket grouping barchan and barchanoid ridge)

[†] May include mounds and other features too small to appear on Landsat imagery.
[‡] Internal structures may show embryo barchan-type with one slipface.
§Dunes controlled by vegetation.

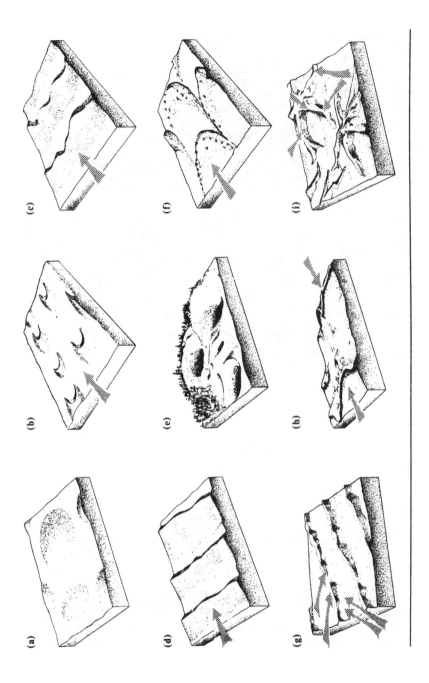

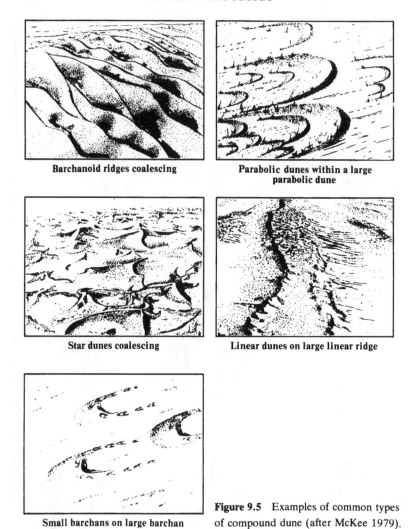

Barchanoid ridges coalescing

Parabolic dunes within a large
parabolic dune

Star dunes coalescing

Linear dunes on large linear ridge

Small barchans on large barchan

Figure 9.5 Examples of common types of compound dune (after McKee 1979).

free dunes of simple, compound and complex types. Simple dunes are individual units; compound dunes are where two or more dunes of the same type occur *en échelon* and complex dunes consist of two or more dunes of different kinds. Figure 9.7 considers dunes related to obstacles.

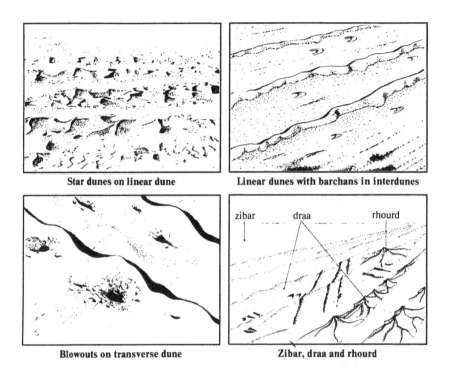

Star dunes on linear dune

Linear dunes with barchans in interdunes

zibar   draa   rhourd

Blowouts on transverse dune

Zibar, draa and rhourd

Figure 9.6 Examples of complex dunes (after McKee 1979, Cooke & Warren 1973).

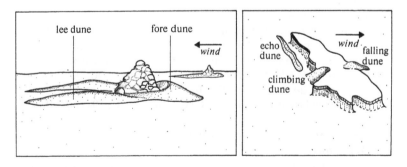

lee dune   fore dune

wind

echo dune   wind   falling dune

climbing dune

Figure 9.7 Classification of dunes related to obstacles (after Cooke & Warren 1973).

10 Coastal processes

This chapter is concerned with the material necessary for the study of estuarine, coastal and shallow-water marine processes. Section 10.1 is a checklist of observations to make under the LEO scheme of the US Corps of Engineers. Section 10.2 consists of a number of items relating to various aspects of water and wave motion and the final section (10.3) gives some material relevant to examination of sediment and salinity. Coastal process measurement techniques are extensively reviewed in Goudie (1981), pp. 247–65.

10.1 The LEO scheme of the US Corps of Engineers

The Littoral Environment Observations (LEO) scheme of the US Corps of Engineers is a widely used procedure by which a full description of coastal processes at a particular point and period of time may be recorded. Full details of it and of procedures to be employed are given by Berg (1968) and Bruno and Hiipakka (1973). A checklist of data to record and simplified recommendations concerning procedures is, however, given below:

(a) *Station identification*
(b) *Date*
(c) *Time* using the 24 h system
(d) *Surf observations*
    (1) *Wave period*: the time in seconds for eleven wave crests to pass some stationary object. Start the timing when the first crest passes the object and stop it on the eleventh. If calm record 0.
    (2) *Breaker height*: based upon observer judgement, to nearest 3 cm.
    (3) *Breaker angle*: the direction from which the waves are approaching the beach with the beach being regarded as 0–180°. If calm record 0.
    (4) *Breaker type*: see pages 189–90.
(e) *Wind observations*
    (1) *Wind speed*: to nearest kilometre per hour from wind meter if available. If this is not available the Beaufort scale (pp. 184–5) may be used to estimate wind speed. If calm record 0.

(2) *Wind direction*: direction from which wind is coming in terms of compass octants.

(f) *Foreshore slope* to the nearest degree – by use of Abney level or inclinometer.

(g) *Width of surf zone* from shore to line of most seaward breakers – in metres.

(h) *Longshore currents*
   (1) '*Dye distance*': dye should be injected just shorewards of the breakers if possible. Driftwood or similar floating objects may also be used. The 'dye distance' is the distance in metres from the shoreline to the point of injection.
   (2) *Current speed*: distance moved by dye cloud or floating object parallel to the shoreline in 1 min – in metres.
   (3) *Current direction*: if when looking seaward the dye has moved to left, record −1; if to right, record +1; if there is no longshore movement, record 0.

(i) *Rip currents*   These are seaward-moving channels of water which return the water piled up by incoming waves; they are fed by feeder currents in which water moves along the shore. If they are present, record their spacing in metres; if absent record 0.

(j) *Beach cusps*   These are semicircular or crescentic cutouts in the beach face. If present, record the distance between the 'horns' of adjacent cusps in metres; if irregular, estimate an average spacing; if absent record 0.

10.2   Aspects of wind and wave motion

This section contains material that is of use in the examination of relationships between aspects of wave geometry, motion and generation. The Beaufort scale for description of wind force is given in Table 10.1. This is followed by a summary of the basic relationships existing between wave geometry and motion; this is included in order to allow preliminary calculations and analysis to be carried out in the field. Breaker types and their field identification are then considered and, finally, material is provided to allow field forecasts and hindcasts concerning wave generation by wind.

Relationships between aspects of wave geometry and motion

Wave velocity or celerity (C) is given by

Table 10.1 The Beaufort scale of wind force.

Beaufort number	General description of wind	Approximate wind speeds ($m\ s^{-1}$)	Approximate wave height (m)	Specification	
				For coastal use	*For inland use*
0	calm	0.0–0.2	0	sea like a mirror	smoke rises vertically
1	light air	0.3–1.5	0.1–0.2	ripples without appearance of scales; no foam crests	wind direction shown by smoke drift but not by wind vanes
2	light breeze	1.6–3.3	0.3–0.5	small wavelets; crests have glassy appearance but do not break	wind felt on face; leaves rustle, ordinary vane moved by wind
3	gentle breeze	3.4–5.4	0.6–1.0	large wavelets; crests begin to break; scattered white horses	leaves and small twigs in constant motion; wind extends light flag
4	moderate breeze	5.5–7.9	1.5	small waves becoming longer; fairly frequent white horses	raises dust and loose paper; small branches are moved
5	fresh breeze	8.0–10.7	2.0	moderate waves; many white horses and chance of some spray	small trees in leaf begin to sway; crested wavelets form on inland water
6	strong breeze	10.8–13.8	3.5	large waves begin to form; white foam crests extensive everywhere and spray probable	large branches in motion; whistling heard in telegraph wires; umbrellas used with difficulty

No.	Description	Wind speed (m/s)	Wave height (m)	Sea conditions	Land conditions
7	near or moderate gale	13.9–17.1	5.0	sea heaps up and white foam from breaking waves begins to be blown in streaks; spindrift begins to be seen	whole trees in motion; inconvenience felt when walking against wind
8	gale or fresh gale	17.2–20.7	7.5	moderate high waves of greater length; foam blown in well marked streaks; edges of crests break into spindrift	breaks twigs off trees; generally impedes progress
9	strong gale	20.8–24.4	9.5	high waves; crests begin to topple and roll over; spray may affect visibility	slight structural damage occurs; chimney pots and slates removed
10	storm or whole gale	24.5–28.4	12.0	very high waves; long overhanging crests, tumbling of sea becomes very heavy and shocking; sea surface takes on white appearance as foam in large patches is blown in very dense streaks	seldom experienced inland; trees uprooted; considerable structural damage occurs
11	violent storm	28.5–32.7	15.0	exceptionally high waves, sea covered by long patches of foam; small and medium-sized ships might be lost to view behind waves for long periods	very rarely experienced; widespread damage
12	hurricane	>32.7	>15	air filled with foam and spray; sea completely white with driving spray; visibility greatly reduced	

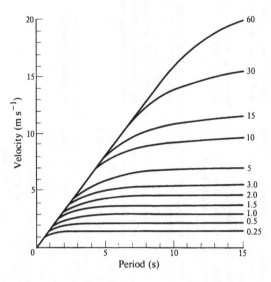

Figure 10.1 Relationship between wave period and velocity for various depths (after O'Brien 1942). Figures on curves represent depths in metres.

$$C = L/T$$

or

$$C = \{(gL/2\pi)\tanh(2\pi d/L)\}^{0.5} = (gT/2\pi)\tanh(2\pi d/L),$$

where L = wavelength (m), T = period (s), d = depth of water (m), g = the gravitational constant (m s^{-2}). A chart showing the relationship between wave velocity, period and depth is given in Figure 10.1, after O'Brien (1942).

The relationship given above may be approximated by

$$C \approx (gL/2\pi)^{0.5} \qquad \text{for deep water } (d/L > 0.5),$$
$$C \approx (gd)^{0.5} \qquad \text{for shallow water } (d/L < 0.05).$$

The deep-water wavelength–period relationship is $L = 1.56T^2$ for L in metres. A more accurate relationship between velocity and other characteristics is

$$C = [(gL/2\pi)\{1 + (\pi^2 H^2/2L^2)\}]^{0.5}.$$

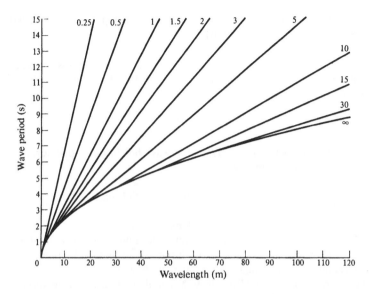

Figure 10.2 Relationship between wave period and wavelength for various depths, for an ideal wave in shallow water. Figures on curves represent depths in metres.

Relationships between wave period and wavelength are shown in Figure 10.2 for various depths; this represents the equation

$$T = \{(2\pi L/g)\coth(2\pi d/L)\}^{0.5}.$$

For deep water the relationship may be regarded as

$$T^2 = 2\pi L/g \approx 0.344L.$$

The wave group velocity (C_g) for a group of waves with velocities C_1 and C_2 may be estimated as $C_g = C_1C_2/(C_1 + C_2)$. If $C_1 \approx C_2$ then $C_g = c/2$, where $c = (C_1 + C_2)/2 = \mathbf{g}T/4$. For shallow depths ($d/L < 0.05$) $C_g \approx C$, where C is mean wave speed. For any depth wave group speed is more exactly given by

$$C_g = 0.5\, C[1 + \{(4\pi d/L)/\sinh(4\pi d/L)\}].$$

188   COASTAL PROCESSES

For further discussion and detail of these relationships see Harvey (1976),
King (1972) and Collins (1976).

Influence of shallowing water and currents on waves   As a wave reaches
shallowing water its height changes (Holmes 1975) according to the relationship

$$H = DH_o,$$

where H = the wave height, H_o = the wave height in deep water, D = a
shoaling coefficient given by

$$D = \tanh(2\pi d/L)[1 + \{(4\pi d/L)/\sinh(4\pi dL)\}]^{-0.5}$$

where d = water depth, L = wavelength.

   In deep water D = 1.0, and it reduces to a minimum of 0.91 at about
d/L_o = 0.15, and then increases until the wave breaks. The relationships
for changing wave height and wavelength with decreasing depth are
shown in Figure 10.3 (Russell & Macmillan 1952). The relationships
between wavelength, wave period and water depth are shown for an ideal
wave in shallow water in Figure 10.2.

   If refraction occurs with shallowing, then its effect upon wave height is
given by

$$H = (b_0/b)^{0.5}DH_o,$$

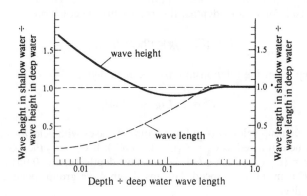

Figure 10.3 Effect of shoaling upon wave height and wavelength (after Russell
& Macmillan 1952).

where b_0 = the distance between wave rays (wave orthogonals) in deep water, b = the distance between the same wave rays in shallow water.

The point at which waves will break is governed by a variety of factors, but in general is given by the semi-empirical relationship

$$H = 0.78d.$$

When the observed value of wave height exceeds this value the wave will break. The influence of a current running with or against a wave is given (Holmes 1975) by

$$L_c/L_o = \{(1 + a)/2\}^2,$$
$$H_c/H_o = \{2/(a^2 + a)\}^{0.5},$$

where the subscript 'c' denotes the current-influenced case and the subscript 'o' denotes the deep-water case. The value of a is given by

$$a = (1 + 4(U/C_o))^{0.5},$$

where U = current velocity, C_o = wave velocity in deep water.

Classification of breaker types

Three main breaker types exist (Fig. 10.4) representing points on a spectrum of types. The type may be predicted by the quantity I (Galvin 1968) given by

$$I = h/(\lambda \tan^2 \beta),$$

where h = wave height in deep water (i.e. still-water depth $\div \lambda > 0.05$), λ = wave length in deep water, β = beach slope. Critical values are indicated in Figure 10.4 and below. Figure 10.4 also gives expectable values of the simpler index h/λ for each type (Patrick & Weigel 1955).

(a) *Spilling breakers* ($I > 4.8$) Commonly on low angle beaches with foam and turbulent water at the crest and eventually the front face of the wave. Spilling commences at the crest when a small tongue of water moves forwards faster than the wave. It evolves finally into a bore. The breaker generally retains shape fairly well.

(b) *Plunging breakers* ($0.09 < I < 4.8$) The front of the wave steepens until vertical. The crest curls and falls into the base of the wave, often

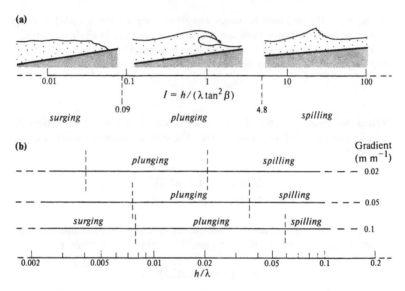

Figure 10.4 Breaker type in relation to indices based upon beach slope, wave height and wavelength. Compiled from data in Galvin (1968) and Patrick and Weigel (1955).

rebounding, and a large sheet-like splash arises where the crest touches down. It deforms very quickly and dramatically.

(c) *Surging breakers* ($I < 0.09$)   Commonly on steep beaches. The front and crest remain relatively smooth and its base rushes up the beach with only slight foam and bubbles. No true breaking occurs and it retains its shape fairly well.

(d) *Collapsing breakers* ($I < 0.09$)   Breakers intermediate between the plunging and surging types may also be recognised. The lower part of the wave front steepens to vertical and curls over as an abbreviated plunging wave. The point where the front face begins to curl is land-wards of and lower than the maximum elevation of the wave.

Forecasting wind-generated sea waves

Figures 10.5–10.8 (after Darbyshire & Draper 1963) may be used to estimate the wave heights and periods likely to result from particular combinations of wind speed, duration and fetch. For oceanic applications use Figures 10.5 and 10.6 and for application to coastal waters use Figures

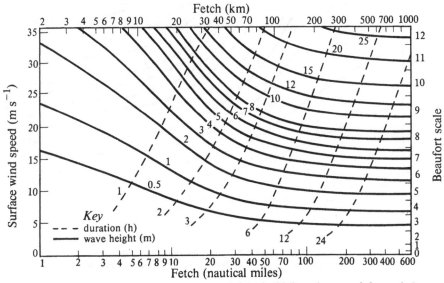

Figure 10.5 Chart for forecasting wave height (solid lines, in metres) from wind speed and fetch, for oceanic waters (after Darbyshire & Draper 1963).

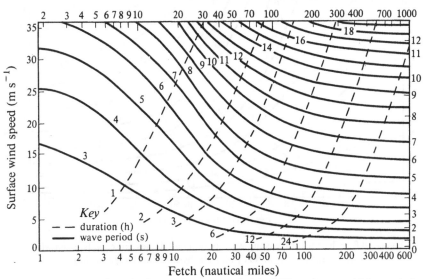

Figure 10.6 Chart for forecasting wave period (solid lines, in seconds) from wind speed and fetch, for oceanic waters (after Darbyshire & Draper 1963).

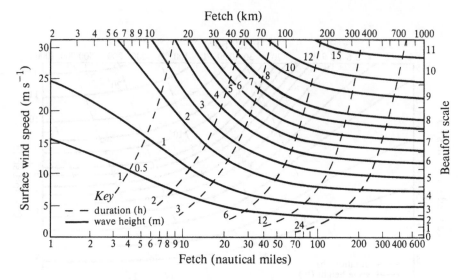

Figure 10.7 Chart for forecasting wave height (solid lines, in metres) from wind speed and fetch, for coastal waters (after Darbyshire and Draper 1963).

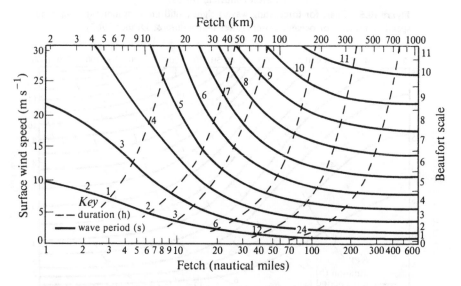

Figure 10.8 Chart for forecasting wave period (solid lines, in seconds) from wind speed and fetch, for coastal waters (after Darbyshire and Draper 1963).

10.7 and 10.8. Figures 10.5–10.8 are used by determining the wind velocity and then entering the appropriate chart at this velocity on the left-hand margin. Figures 10.5 and 10.7 are used for forecasting wave heights, Figures 10.6 and 10.8 for wave periods. Having entered at the appropriate velocity value in the appropriate chart the value is followed across the chart until it meets either the duration value or the fetch. The desired wave height or period is then read off from the bottom scale of the figure at this point. The first limiting value is always used, whether it be for duration or fetch. The figures obtained are the maximum wave height (H_{max}) and the significant wave period (T_s). This is the period that would appear on a typical wave record of 10 min duration and containing c. 100 waves. The significant wave height (H_s) is the mean height of the highest one-third of waves on the record and it may be estimated from

$$H_s = H_{max}/1.60.$$

The most probable height of the maximum wave in the storm (H_p) may be estimated from

$$H_p = kH_{max},$$

where k is a coefficient which may be derived (Longuet-Higgins 1952) from Figure 10.9.

The average wave height (H_{av}) is given by

$$H_{av} = 0.625\ H_s$$

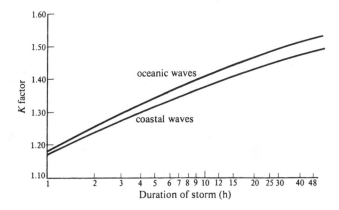

Figure 10.9 Derivation of k factor for estimation of most probable height of maximum wave in storm from maximum wave height (after Longuet-Higgins 1952).

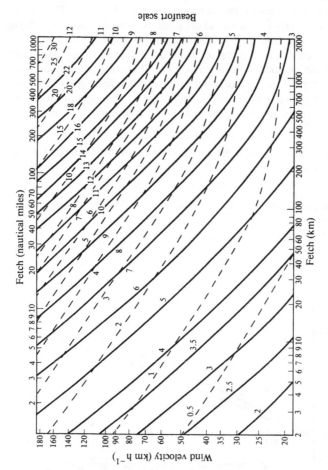

Figure 10.10 Chart for deep-water forecasting of deep-water wave heights (dotted lines, in metres) and period (solid lines, in seconds), as functions of wind velocity and fetch (after Bretschneider 1966).

and the mean height of the highest 10% of the waves ($H_{0.1}$) by

$$H_{0.1} = 1.27 \, H_s .$$

An alternative chart for wave forecasting in oceanic waters is given in Figure 10.10 after Bretschneider (1966). This may be preferred to the Darbyshire and Draper charts, although no guidelines can be offered as to when each chart is most useful since many authorities disagree as to which method is most accurate (King 1972) and under which conditions each should be used. Figure 10.11, also after Bretschneider (1966), is intended for application to shallow water (<30 m) of constant depth and may therefore be of use in both marine and limnological studies.

10.3 Sediment and salinity

This section contains material allowing an estimate of marine bedload transport to be made in the field and a temperature–salinity diagram allowing observations on salinity to be subject to preliminary examination. For formulae relating to longshore current velocities and sediment

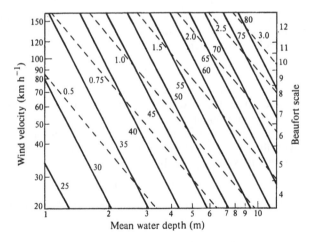

Figure 10.11 Chart for forecasting wave period (solid lines, in seconds) and significant height (dotted lines, in metres) in terms of wind velocity and mean water depth. This is for application to shallow water of constant depth (after Bretschneider 1966).

transport, reference should be made to Galvin (1967), King (1972) and Inman and Bagnold (1963); for current velocities necessary to initiate sediment movement and types of sediment transport in relation to velocity and particle size see Open University (1978). For information concerning bedforms see Clifton (1976).

Estimation of bedload transport for shallow marine environments

Bedload sediment transport in shallow marine environments may be estimated from measurements of sediment texture and mean velocity of current flow 1 m above the bed. The equations of Bagnold (1963) provide the basis of the method, as modified in the light of further field observations by Sternberg (1972). The method was presented in this form by Sternberg. The specific steps in the estimation procedure are as follows:

Step 1: Estimate the boundary shear stress (τ_0) from velocity measurements. If the velocity profile is known, then

$$\tau_0 = \rho[(\bar{U}_{z2} - \bar{U}_{z1})/5.75(\log z_2 - \log z_1)]^2,$$

where ρ = fluid density, $\bar{U}_{z1}, \bar{U}_{z2}$ = mean velocities at heights z_1 and z_2 from the bed. If the mean velocity at 1 m above the bed ($\bar{U}_{100}$) is known, then

$$\tau_0 = 3 \times 10^{-3}\rho\bar{U}_{100}^2.$$

Step 2: Estimate mean sediment diameter (d). Using this, estimate from Figure 10.12 the critical shear stress (τ_c) required to initiate sediment movement.

Step 3: Calculate the excess shear stress (τ_e) and use this in conjunction with the value of d in Figure 10.13 to estimate the value of the coefficient K. τ_e is given by

$$\tau_e = (\tau_0 - \tau_c)/\tau_c.$$

Step 4: Calculate the estimate of bedload transport rate (j) from

$$j = K\rho U_*^3/gh,$$

where g = the gravitational constant, U_*^3 = the friction velocity = $5.47 \times 10^{-2}U_{100}$, $h = (\rho_s - \rho)/\rho_s$, where ρ_s is the sediment density. This calculation can be performed by the use of the nomogram in Figure 10.14.

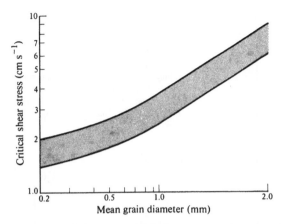

Figure 10.12 Relationship between critical shear stress and mean grain size. The stippled band represents the broad area within which values have been reported (after Inman 1963, and employing data sources quoted therein).

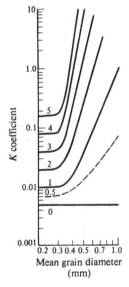

Figure 10.13 Relationship between the K coefficient in the bedload estimation equation and mean grain size, for different values of excess shear stress (after Sternberg 1972). Numbers on curves represent values of excess shear stress.

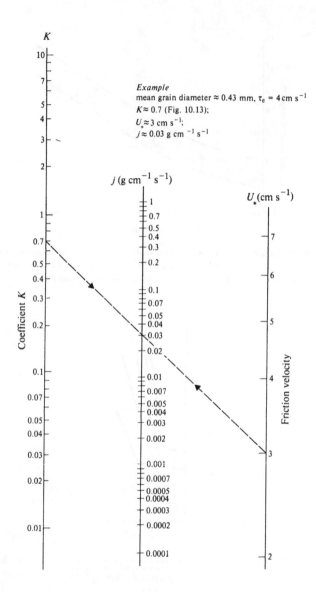

Figure 10.14 Nomogram for evaluation of bedload transport, from measurements of the critical velocity and the coefficient K derived by the method described in the text (Sternberg 1972).

This technique has the following limitations:
(a) In a tidal current, flow should be in an accelerating or relatively steady part of the tidal cycle.
(b) It is applicable for mean sediment sizes ranging from approximately 0.20 mm to 2 mm.
(c) Sediment transport should be largely as bedload, not as suspended sediment, and therefore is applicable to flow conditions given by Figure 10.15.

Temperature–salinity diagram

Figure 10.16 is a T–S diagram on which values of temperature and salinity for water samples may be plotted. It may be used for water-mass recognition, for ascertaining the stability of water stratification, and for investigation of water mixing (Mamayev 1975). The lines plotted in the figure portray equal values of a parameter σ, defined by $\sigma = 10^3(\rho_w - 1)$, where ρ_w is the density of water. For most purposes the small effect of pressure on water density can be disregarded and the values of σ shown are those corresponding to atmospheric pressure. Negative temperatures are included because the freezing point of saline water is below 0 °C, as shown on the figure. The chart covers a wide range of temperature and salinity values and may be used for data from coastal, marine and estuarine environments.

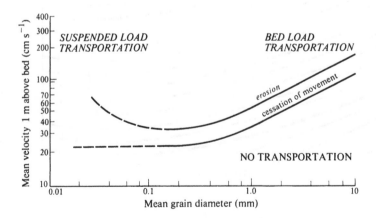

Figure 10.15 Relationship between grain size, flow velocity and type of sediment movement, assuming a sediment density of 2.65 g cm^{-5} (after Sundborg 1967).

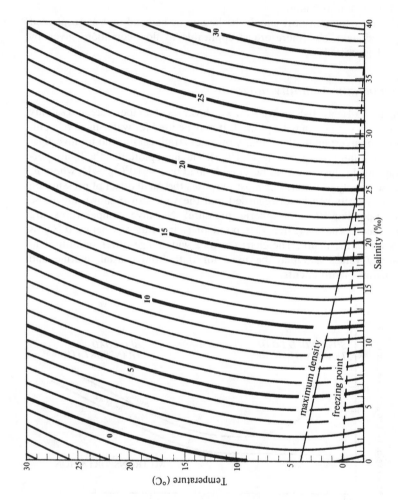

Figure 10.16 Temperature–salinity diagram. The parameter σ is explained in the text.

11 Slope processes

This chapter is concerned with the examination of slope processes. For a fuller review of appropriate methods see Goudie (1981), pp. 156–80. The first section contains guidance in the classification of mass-movement processes. Section 11.2 gives basic relationships for preliminary examination of slope stability, and the final section gives field methods for estimation of particular material properties relevant to slope stability studies.

11.1 Identification and classification of slope mass-movement processes

Classification of mass-movement processes may be attempted in a great variety of ways and according to many different criteria. Figure 11.1

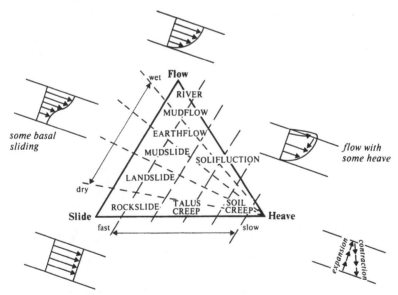

Figure 11.1 Classification of mass-movement processes. Processes are classified according to the relative importance of the three types of motion. Schematic velocity profiles of these types of motion are shown in the figure (Carson & Kirkby 1972).

shows a simple classification of processes according to the types of motion involved. This emphasises the way in which more than one type of motion is usually involved in the majority of processes and how they therefore grade into one another. It excludes processes resulting from falling motions. A simpler classification of slope failures based upon structural conditions is given in Figure 11.2 and Table 11.1.

Landslides and landslips are further classified in Figure 11.3. These are features which occur 'primarily as a result of shear failure at the boundaries of the moving mass' (Brunsden 1971) and include both sliding and flowing motions.

Falls may be divided into rockfalls and debris falls. They are features that involve movement of the mass as a whole resulting from material failure and involving predominantly vertical motion with little outward motion except by bouncing.

The classifications suggested above represent only some of the many possible approaches to slope process classification. For more detailed

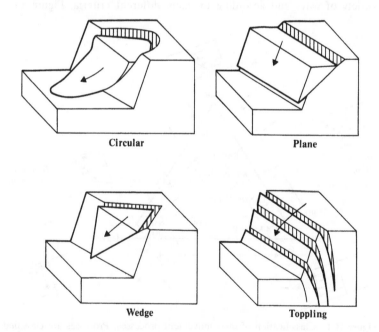

Figure 11.2 A classification of types of slope failure in terms of the structural conditions. Failure planes are shown stippled (Hoek & Bray 1977). These are described in Table 11.1.

Table 11.1 Characteristics of main types of slope failure in relation to structural conditions (after Hoek & Bray 1977). The types of slope failure are illustrated in Figure 11.2.

Type	Characteristics
circular failure	in overburden soil, waste material or heavily fractured rock with no identifiable structural pattern
plane failure	in rock with highly ordered structure, e.g. slate
wedge failure	where failure occurs on two intersecting discontinuities
toppling failure	in hard rock in which columnar structures are separated by steeply dipping discontinuities

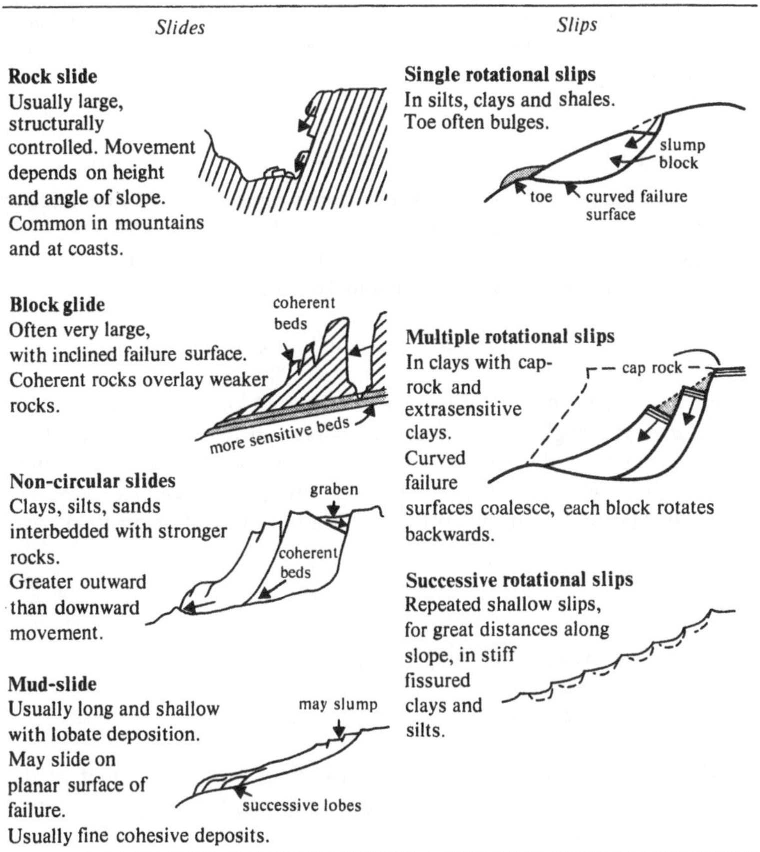

Slides   *Slips*

Rock slide
Usually large, structurally controlled. Movement depends on height and angle of slope. Common in mountains and at coasts.

Single rotational slips
In silts, clays and shales. Toe often bulges.

slump block

toe   curved failure surface

Block glide
Often very large, with inclined failure surface. Coherent rocks overlay weaker rocks.

coherent beds

more sensitive beds

Multiple rotational slips
In clays with cap-rock and extrasensitive clays. Curved failure surfaces coalesce, each block rotates backwards.

cap rock

Non-circular slides
Clays, silts, sands interbedded with stronger rocks. Greater outward than downward movement.

graben

coherent beds

Successive rotational slips
Repeated shallow slips, for great distances along slope, in stiff fissured clays and silts.

Mud-slide
Usually long and shallow with lobate deposition. May slide on planar surface of failure. Usually fine cohesive deposits.

may slump

successive lobes

Figure 11.3 A classification of landslides and landslips (Brunsden 1971).

classifications, not all of which may be readily applied in the field, see Young (1972), Hutchinson (1968), de Quervain (1966), Brunsden (1971), Sharpe (1938) and Varnes (1958).

11.2 Slope stability

This section includes relationships which allow an approximate analysis of slope stability in the field. However, it includes only the simplest forms of analysis, which are applicable only to average conditions and materials. For specific detailed analyses and stability calculations reference should be made to specialised texts such as Carson and Kirkby (1972), Hoek and Bray (1977), Chowdhury (1978), Terzaghi (1943) and Terzaghi and Peck (1967).

Rock slopes

For more rigorous descriptions of rock slopes reference should be made to standard texts (e.g. Carson & Kirkby 1972, Hoek & Bray 1977).

A general equation relating critical slope height (H_c) to properties of the material and slope angle (Culmann 1866) is

$$H_c = 2c \sin i / \gamma \sin(i - a)(\sin a - \cos a \tan \varphi) \, ,$$

where c = cohesion, γ = unit weight of the rock or soil, i = the angle of slope, a = the angle of the potential failure surface, φ = the angle of internal friction of the material. Methods for estimation of φ in the field are given on page 208. Typical values of c and γ are given in Tables 11.3, 4, 5 & 6. The angle a at the critical condition is given (Carson 1971) by

$$a = 0.5(i + \varphi)$$

and therefore the equation may be regarded as

$$H_c = 4c \sin i \cos \varphi / \gamma \{1 - \cos(i - \varphi)\} \, .$$

When the slope has an inclination less than or equal to that of the potential failure plane, the critical height is infinite. If the angle of internal friction is zero, then

$$H_c = 4c \sin i / \gamma (1 - \cos i) \, ,$$

and if in addition the slope is vertical,

$$H_c = 4c/\gamma \, .$$

In the case where tension cracks develop in the banks prior to failure the critical height (H'_c) is given (Terzaghi 1943) by

$$H'_c = H_c - z \, ,$$

where z = the depth of the tension cracks. When the slope is a vertical cliff, the general formula above reduces to

$$H_c = 4c \cos \varphi/\gamma(1 - \sin \varphi) = \frac{4c \, \tan(45 + \varphi/2)}{\gamma}$$

if no tension cracks occur. If tension cracks occur, the appropriate relationship for critical height (H'_c) of a slope subject to slab failure is given by Lohnes and Handy (1968) as

$$H'_c = \{4c/\gamma(\cos \varphi - 2 \cos^2 a \, \tan \varphi)\} - z \, ,$$

where $a = 45 + \varphi/2$. This may also be written (Terzaghi 1943) as

$$H'_c = \frac{4c \, \tan(45 + \varphi/2)}{\gamma} - z.$$

If the material has zero tensile strength, this simplifies to

$$H'_c = \frac{2c \, \tan(45 + \varphi/2)}{\gamma}$$

The critical height of an unweathered rock mass is given approximately (Terzaghi 1943) by

$$H'_c = q/\gamma \, ,$$

where q = the unconfined compressive strength of the rock. The value given by this equation is practically never achieved in reality, however, because the vertical height is reduced by the presence of mechanical defects such as joints, fissures or faults, and the relationship of Lohnes and Handy (1968) above is usually thought to be more realistic.

Instability in soils and waste-mantled slopes

For shallow slides in which the failure surface is approximately planar and parallel to the surface, in which it may be assumed that the water table is at the surface and the soil is a layer of infinite extent relative to its depth, the factor of safety (F_s) is given by

$$F_s = s/\tau \, ,$$

where s = the shear strength along the base of a column of soil and τ = the shear stress along the failure surface. These may be estimated from

$$s = c' + z \cos^2\theta(\gamma - \gamma_w)\tan\varphi \, ,$$
$$\tau = \gamma z \sin \theta \cos \theta,$$

where c' = cohesion (effective stress value) ≈ 0 if deeply fissured, z = depth to the failure plane, θ = angle of slope, γ = bulk unit weight of the soil, γ_w = unit weight of water, φ' = angle of internal friction (effective stress value). At the moment of failure $F_s = 1.0$.

For deep-seated slips reference may be made to Figure 11.4. This employs a dimensionless slope stability number, N_s, given by

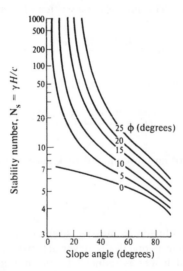

Figure 11.4 Stability chart for deep-seated slips. φ = the angle of internal friction of the material; the use of the stability number is explained in the text (Scott 1963).

$$N_s = \gamma H/c \, ,$$

where γ = unit weight of material, H = the depth of the cut forming the slope, c = cohesion, φ = the angle of internal friction. Figure 11.4 may be used with values of N_s estimated from material properties and values of φ to estimate the maximum stable slope angle in a particular situation.

The methods given above for shallow and deep-seated slips represent only very simple methods which may be useful for rapid field estimates. For more thorough analytical methods reference should be made to appropriate texts (e.g. Carson & Kirkby 1972, Bowles 1979, Hoek & Bray 1977, Terzaghi & Peck 1967). More comprehensive stability charts are presented by Bishop and Morgenstern (1960) and Cousins (1978), and are reviewed by Chowdhury (1978).

11.3 Field estimation of material properties significant to slope stability

This section incorporates methods for estimation of important properties for use in slope stability analyses as outlined in the previous section. The data and relationships given, however, are only applicable in an average sense and should only be used to derive field estimates; for more rigorous analyses the significant properties must be determined by the appropriate laboratory analytical techniques (BS 1377 1975), as described in Goudie (1981), pp. 79–131.

Determination of shear strength, cohesion and the angle of internal friction in slope stability studies

These properties may be defined (Coulomb 1776) by

$$s = c + \sigma \tan \varphi$$

where s = shear strength, c = cohesion, σ = normal stress, φ = angle of internal friction, although in reality c is not independent of σ (Hvorslev 1960), and is better thought of as effective cohesion; the precise determination of s, c and φ requires the use of direct shear test apparatus in a laboratory. Approximate estimates of them may, however, be arrived at in the field by one of the following procedures:

(a) For granular material in its loosest state the angle of internal friction approximates, according to Skempton (1945), to its angle of repose when poured on a level surface in the shape of a cone.

(b) For sandy materials Table 11.2 gives a set of generalised relationships suggested by Kirkpatrick (1965) between mean particle size, void ratio and the angle of internal friction.

(c) For soils Figure 11.5 gives a generalised relationship between percentage clay and the angle of internal friction (Skempton 1964).

(d) For talus and gravelly materials, typical values are given in Table 11.3.

(e) For mixed talus and colluvium material typical values are given in Table 11.4.

Table 11.2 Approximate angle of internal friction (degrees) for sandy soils (after Kirkpatrick 1965).

Mean particle size (mm)	Void ratio		
	0.54	0.60	0.67
0.25	43	39½	37
0.5	42½	39	36½
1.0	41	38	35
1.5	40	37	34
1.75	39½	36	33½

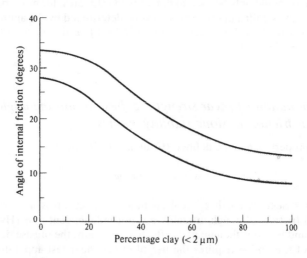

Figure 11.5 Relationship between residual angle of internal friction and percentage clay content (after Skempton 1964).

(f) For residual soils typical values are given in Table 11.5.
(g) For intact rocks typical values are given in Table 11.6. (Note that the values given in Table 11.6 refer to values for intact rock and that these may not be directly applicable to slope stability in well jointed rock masses. For hard massive rocks with a random joint pattern Terzaghi (1943) suggests a value of angle of internal friction c. 70° to be appropriate and Carson and Kirkby (1972) suggest that a range of values from 45° to 75° may well apply to fractured rock systems.)
(h) Approximate field tests for determination of shear strength are given by Table 11.7.

(a), (b) and (c) may be regarded as very generalised relationships and should be used as the basis for only very approximate estimates; in certain cases values may differ widely from those suggested.

A very approximate estimate of unconfined compressive strength may be made in the field by one of the following methods:

Table 11.3 Shear strength, cohesion and angle of internal friction for talus materials (gravel fraction > 90%). Data from Carson and Kirkby (1972) and Hoek and Bray (1977).

Material	Median particle size (cm)	Void ratio	Cohesion (kgf cm^{-2})	Maximum angle of internal friction (degrees)[†]	Shear strength (kgf cm^{-2})
A. Reported values for actual examples					
Carboniferous limestone	2	0.83	0	53	0–0.2
basalt (a)	1	0.30	0	45	0–10
(b)	1	0.30	8	30	20–50
granitic gneiss (a)	5.5	0.32	0	36	0–10
(b)	11	0.62	0	31	0–10
(c)	5.5	0.32	6	22	30–50
(d)	11	0.62	7	18	30–50
B. General ranges of values for blasted/broken rock					
limestone				35–40	
basalt				40–50	
chalk				30–40	
granite				45–50	
sandstone				35–45	
shale				30–35	

[†]Use higher values for more angular material.

Table 11.4 Shear strength, and angle of internal friction for mixed talus and colluvium material. Data from Carson and Kirkby (1972), Hoek and Bray (1977) and Bowles (1979).

Material	Gravel fraction (%)	Void ratio	Maximum angle of internal friction (degrees)[†]	Shear strength (kgf cm^{-2})
A. Reported value for actual examples				
alluvium	70	0.3	41–44	3.5–24
river deposit	20–60	0.4–0.55	45–48	0–1.4
quarry rubble	60–80	0.3	39–44	0–1.4
silty, sandy gravel		0.15	45	0–7
glacial till			37	0–7
slaty-greywacke	85	0.45–0.7	43	
Exmoor slate	57	0.5	44	0–0.2
shale grit	40–60	0.4–0.5	43	0–0.2
fluvial, glacial, talus, etc.	60–80		37	0–4
B. General range of values				
gravel, uniform grain size			34–37	
sand and gravel, mixed grain size			45–48	
loose sandy gravel	0.44–0.73		30–38	
dense sandy gravel	0.18–0.44		36–45	
loose gravel	0.44–0.62		32–36	
dense gravel	0.30–0.44		35–50	

†Use higher values for more angular material.

(a) Under undrained conditions it may be taken as equivalent to undrained shear strength, which may be estimated from Table 11.7. Under these conditions undrained compressive strength also equals twice the cohesion, which may be estimated from Tables 11.3, 5 and 6.

(b) For rocks Figure 11.6 indicates the approximate range of expectable values.

(c) Unconfined compressive strength may be related to the standard penetration test. The number of blows per metre required in this test in relation to approximate compressive strengths is given in Table 11.7.

(d) Figure 6.1 may be used to relate Schmidt hammer impact values to approximate values of unconfined compressive strength.

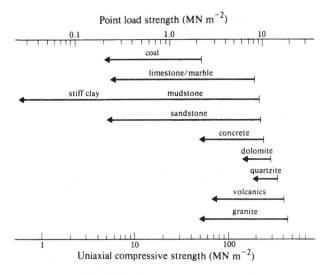

Figure 11.6 The correlation between point load strength and uniaxial compressive strength, and typical ranges of values for some common rock types (Roberts 1977).

Classification of clays

Clays may be classified for many slope studies according to two criteria: (a) consistence – soft or stiff; (b) fissuring – intact or fissured. This classification applies to clays in their saturated state.

(a) The distinction between soft and stiff is based upon the liquidity index (see Goudie 1981) as follows:

soft	liquidity index ≥ 0.5
stiff	liquidity index < 0.5, and typically $c.\ 0.0$.

Soft clays are usually normally consolidated or lightly over-consolidated, whereas stiff clays are usually heavily over-consolidated. Cohesion is $c.\ 0.0$ for soft clays whilst for stiff clays it is larger (e.g. $c.\ 9.5$ kPa). Strength properties for soft and stiff clays also differ (Table 11.5). At their upper limit stiff clays grade to hard clays or 'clay shales' with undrained shear strengths of greater than 200 kPa. These can still be worked with water to form a paste, whereas shales and mudstones are more indurated and they resist disintegration in water. The strength and cohesion properties men-

Table 11.5 Shear strength, cohesion and angle of internal friction for residual soils, clays and sands. Data from Carson and Kirkby (1972), Hoek and Bray (1977) and Bowles (1979).

Parent material	Clay fraction (<2 μm) (%)	Void ratio	Cohesion ($kgf\ cm^{-2}$)	Maximum angle of internal friction (degrees) (residual values bracketed)†	Shear strength ($kgf\ cm^{-2}$)
A. Reported values for actual examples					
sand; England	0	0.5–0.7	0	33–43	
molsand; Belgium	0			35–50	
slate; England	0	0.6	0	42 (42)	0.3
sandstone; England	0	0.7	0	36 (36)	1.5
granite–gneiss; Brazil	5		0	42	
granite; Hong Kong	5	0.4–0.6		36–37	
sandstone; Brazil	6–22		0.1–0.5	30–33	
silt	10		0	32–36	
mudstone; England	36		0.1	25 (19)	2.1
gneiss; Brazil	6–42		0.2–0.4	26–29	
basalt; Brazil	10–47		0.4–0.5	29–30	
London clay	55		0.15	20 (16)	2.8

B. General range of values for clays and sands

loose sand, uniform grain size					28–34
loose coarse sand	50–73				32–38
dense sand, uniform grain size					32–40
dense coarse sand	33–50				35–48
loose sand, mixed grain size					34–40
loose clayey sand	58–86				28–32
dense sand, mixed grain size					38–46
dense clayey sand		40–58			35–40
soft bentonite clay			0.1–0.2		7–13
very soft organic clay			0.1–0.3		12–16
soft, slightly organic clay			0.2–0.5		22–27
soft glacial clay			0.3–0.7		27–32
stiff glacial clay			0.7–1.5		30–32
glacial till			1.5–2.5		32–35
loose silty sand				0.68–1.05	28–32
dense silty sand				0.49–0.68	32–38
loose fine sand				0.40–0.86	27–33
dense fine sand				0.44–0.68	33–39
loose gravelly sand				0.44–0.73	30–38
dense gravelly sand				0.16–0.44	36–50
loose silt				0.68–0.86	20–30
dense silt				0.49–0.68	25–32

†Use higher values for more angular material.

tioned above normally require laboratory determination, but for approximation methods see page 208.

(b) Intact clays are free from joints and fissures, whereas fissured clays have a network of discontinuities of various types, e.g. laminations, fissures, joints, faults or slickensides.

Each group of clays may be further divided according to its liquid limit index; Skempton (1964) suggests the following:

Type	Liquid limit index (%)
sandy or silty clays	<30
clays of low plasticity	30–50
clays of medium plasticity	50–90
clays of high plasticity	>90

A further index of clay type is sensitivity (S_t) given by

S_t = undisturbed undrained strength ÷ remoulded undrained strength .

A suggested classification is

Sensitivity	Type
<2	insensitive
2–4	medium sensitivity
4–8	sensitive
8–16	very sensitive
>16	quickclay

Table 11.6 Cohesion and angle of internal friction for intact rock. Data from Carson and Kirkby (1972) and Hoek and Bray (1977).

Material	Cohesion $(kgf\,cm^{-2})$	Angle of internal friction (degrees)	Normal stress $(kgf\,cm^{-2})$
A. Reported values for actual examples			
chalk	9	21	0–70
sandstone (a)	350	44	0–350
(b)	42–50	48–50	210–2100
siltstone	210	29	0–350
limestone	35–350	37–58	210–2100
granite	97–406	51–58	700–2800
B. General range of values			
hard igneous rocks	360–560	35–45	
metamorphic rocks	200–410	30–40	
hard sedimentary rocks	100–300	35–45	
soft sedimentary rocks	10–200	25–35	

Table 11.7 Approximate field identification of soil or sediment consistency and strength (after Bowles 1979).

Consistency	Undrained shear strength (unconfined compressive strength) (kPa)	Field identification	Standard penetration test (blows) per metre)
very soft	<25	easily penetrated several centimetres by fist	<7
soft	25–50	easily penetrated several centimetres by thumb	7–13
medium	50–100	moderate effort required to penetrate several centimetres with thumb	13–26
stiff	100–200	readily indented by thumb	26–52
very stiff	200–400	readily indented by thumbnail	52–105
hard	>400	difficult to indent with thumbnail	>105

12 Sampling

This chapter contains four elements. The first (Sec. 12.1) is material of a statistical nature for random sampling of geomorphological populations. This is followed (Sec. 12.2) by recommendations concerning sample sizes necessary for determination of particle size distribution and then recommendations and procedures for rock, soil and fossil collection (Sec. 12.3), including materials for radiocarbon dating. The final section gives recommendations concerning water sampling.

12.1 Random sampling

Each single digit in Table 12.1 is an independent sample from a population in which the digits 0–9 are equally likely, each having a probability of occurrence of 0.1. They may be used for drawing random samples from populations of objects such as pebbles or for generating random co-ordinates at which measurements such as slope angle may be made. Figure 12.1 (Sen 1978) allows a preliminary sample of measurements to be used to estimate the final sample size necessary to establish a population mean within a given confidence limit. The figure is used as follows:

(1) Take a small sample and choose a probability level. In most geomorphological applications 0.95 would be appropriate.
(2) Determine the standard deviation of the measured characteristic, such as pebble size or roundness.
(3) Decide the confidence limits within which it is wished to determine the population mean in data units above and below the mean. This must be decided according to the purposes of the individual study.
(4) Express the standard deviation of the initial sample in multiples of this confidence limit, i.e. divide the standard deviation by the confidence limit.
(5) Follow the appropriate curve as determined by initial sample size and probability level (right margin) to its intersection with the appropriate value of standard deviation (left margin), and read off the final sample size at this point from the bottom of the chart.

Table 12.1   Random numbers. Each digit in this table is a random number, the digits 0–9 each having equal probabilities of occurrence.

72328	01405	95661	83420	78000	03584	89256	48148	54441	15006	75844
50353	53906	43341	96425	80011	80602	51820	08747	05310	09154	85027
40095	55564	77525	64382	13597	58097	02658	23665	47217	61670	67507
00672	32089	39260	21678	37862	51031	02383	94270	24767	62636	56282
85373	82265	01452	02664	95827	63541	57748	37788	96641	13074	59851
60211	29095	08662	76489	13400	80308	33724	66220	23324	65317	03088
63654	67504	26671	55107	28320	06115	20688	11712	78786	66667	09120
33219	69719	37069	06584	89301	23188	47587	97659	16735	85383	40863
19059	48330	18547	05172	25732	48938	25173	56544	18643	46562	56058
28028	91527	34880	65788	77156	40212	16872	30322	47825	78698	89286
95262	87260	10265	28031	15641	49008	44362	68420	44736	80945	54083
68371	81282	26262	66386	11234	40456	77640	20932	55145	09798	61267
02517	72647	53029	21456	12820	17858	53396	72542	56587	61588	30946
55912	42538	65268	87143	85820	73838	73762	84002	56484	19486	27427
52711	65041	93170	13614	17029	18270	75218	30374	09200	22651	76392
86583	45872	33471	45101	67867	21767	74746	37426	23953	43742	60724
04562	14705	32767	85383	43730	74666	62489	24164	42771	56411	55458
28192	88637	15307	88580	78652	79156	39632	62864	84734	10681	57011
47173	54956	75299	10605	07643	70394	56712	20736	12631	64566	81016
57943	37570	62821	50387	53126	38468	21570	61830	21384	86599	07684
33235	25578	59405	24712	14565	98587	94862	61568	52375	51258	03932
57567	66833	39346	20099	68204	09323	94646	05511	29860	57563	27556
12043	42767	17828	73918	05445	16253	10615	04136	82898	31182	07834
30287	81277	71234	51435	35285	21856	03944	05070	52557	86600	66672
54164	48823	28124	38875	08937	51087	31415	21947	67216	25701	10042
36307	71783	99230	88528	35911	28077	31027	51213	37553	00330	12540
28660	87182	65085	93923	60378	48458	24842	71311	44193	39467	37422
07107	55856	42467	46407	21451	76500	18655	88052	41106	19538	64092
11026	67355	54195	91122	55017	41382	84028	05463	34210	68635	44965
69610	54521	82653	64520	96323	92538	31232	51428	70460	78712	01378
72836	66023	50517	74013	85441	78707	00286	65630	11430	60014	26046
73090	58675	26463	56635	96127	10657	80026	40735	90560	27350	00610
47466	54451	81515	38197	24982	38376	91496	05060	51789	16501	54418
18207	41438	55220	55808	54785	47129	06123	87318	19685	53824	51234
52982	03033	01617	63932	36097	86389	21048	33319	98684	48544	41435

12.2   Sample amounts necessary for determination of particle size distribution

For coarse-grained sediments, Figure 12.2 (Jansen 1979) and Table 12.2 (BS 812 1975, BS 3681 1973) give an indication of the minimum sample sizes desirable for determination of particle size distributions. Figure 12.2 was derived theoretically for application to fluviatile sediments and it recommends much larger samples than Table 12.2, which is intended for application to well sorted sediments. For finer-grained sediments no minimum size of sample can be given. However, the maximum mass of the entire sample should not be so great as to cause clogging of sieves. In practice 100–150 g is appropriate for coarser sands and 40–60 g for fine sands. More precise guidelines as to the appropriate amounts to be sieved in the laboratory are given in BS 812 (1975) and BS 3681 (1973). In

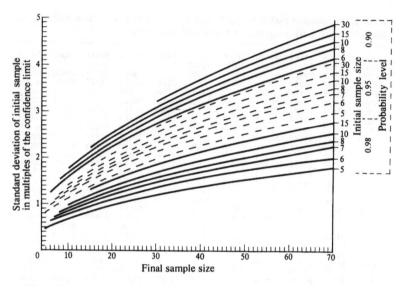

Figure 12.1 Nomogram for estimation of sample size necessary to establish a population mean within a given confidence limit (based upon Sen 1978).

Table 12.2 Minimum sample sizes required for particle size distribution determination by sieving (after BS 812 1975, and BS 3681 1973). Similar values are also appropriate for sedimentation and decantation methods.

Nominal size of material (mm)	Minimum mass of sample (kg)	
	High density†	Low density†
63	50	—
50	35	—
40	15	7.5
28	5	3
20	2	1
14	1	0.5
10	0.5	0.25
6	0.2	0.1
5	0.2	0.1
3	0.2	0.1
3	0.1	—

†Low density materials have bulk densities <1200 kg m^{-3} for materials of less than 5 mm nominal size, or 950 kg m^{-3} for coarser materials.

application of the guidelines given above allowance should, of course, be made for duplicate analyses if required.

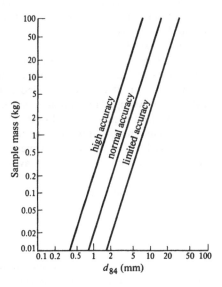

Figure 12.2 Minimum sample size necessary for particle size distribution determination of coarse-grained fluviatile sediments. d_{84} is the value of sediment size which 84% of the deposit is coarser than, and is estimated from a preliminary sample (Jansen 1979).

12.3   Rock, fossil and soil sampling, collection and transport

General recommendations concerning procedures for sampling and collection of rocks, fossils and soils are given in this section, after Compton (1962) and British Museum (1970). Transport and storage of specimens is also considered.

Rocks and soil

Samples should be chosen carefully to represent unweathered exposures. The geographic and stratigraphic locations, the name of the collector and the date should be recorded. Where directional relations are important the sample should be marked to indicate its orientation. All samples should preferably be stored in plastic containers with labels inside and outside. The samples should not be marked with paint if intended for chemical analysis. Powdered soils and rocks should be kept in airtight plastic or glass containers. Rock samples may be individually wrapped in newspaper or similar material and packed in canvas, tough manilla or plastic bags. These should be packed with paper or straw, but not sawdust, in small, strong, wooden boxes for transport.

A screw auger should be used for sampling soil for soil moisture determination because this does not incorporate stones in the sample to the same extent as a core sampler. It is recommended (Curtis & Trudgill 1975) that samples would be wrapped in aluminium foil and be marked by indelible felt-tipped pens or on paper enclosed in the foil or attached by paper clips. Self-sealing tins are an alternative to foil and, for highest accuracy, glass tubes corked and sealed with petroleum jelly and scotch tape should be used. Prolonged storage should be avoided but, if necessary, samples should be stored at as low a temperature as is possible without freezing in glass tubes sealed as above.

Fossils

Fossils should preferably be collected from unweathered exposures, retaining some of the original rock on the specimen if possible. An accompanying label should bear appropriate geological data such as the position and attitude of fossils in the sequence. Damp specimens should be packed wet and allowed to dry slowly. Friable fossils should be hardened in the field by spraying or coating with clear plastic or shellac. Dry specimens should be treated with a thin solution (10% by weight) of 'Butvar 98' in isopropyl alcohol, or methylated spirits to which 10% by volume of diacetone alcohol has been added. An alternative is a solution of button shellac in methylated spirits (knotting) diluted with methylated spirits as necessary. Damp specimens should be treated with an emulsion of polyvinyl acetate. This can be made from a 20% weight to volume solution of polyvinyl acetate in toluene and up to 25% by volume of a detergent. Fossil leaves should not, however, be hardened. Large friable specimens such as vertebrate skeletons should be hardened as above, then covered with thin, wet paper followed by plaster of Paris, supplemented by sacking soaked in plaster of Paris if necessary. Moderately delicate bones can be transported by binding with long strips of fabric soaked in flour paste, strengthened with wood or iron bars bound in as necessary. Small delicate specimens should be packed in cotton wool or small tins or boxes. Others should be wrapped in tissue, then newspaper, or just newspaper, and should be packed tightly to prevent rubbing, in containers as for rock specimens.

Collection and storage of samples for radiocarbon dating

Laboratories able to conduct radiocarbon assays for dating purposes are listed in the third issue each year of the journal *Radiocarbon*. Each laboratory has its own detailed requirements concerning material submitted for assay. Since there are now many laboratories only very general guidelines (after Jones & Cundill 1978, Worsley 1981) can be given here as to the procedures which should be adopted in collection and storage of materials. Most laboratories have an order of preference for material ranging from charcoal, which is most suitable, through wood, peat and bone to shell, which is the least suitable. A reliable date can be produced from *c.* 5 g of carbon. In a carbon-rich deposit this may require *c.* 30 g of sediment. If the sediment's carbon content is low, as much as 100 g may be necessary (Table 12.3). Enough material to allow a duplicate sample should always be collected.

The provenance of material must be precisely delimited; the sample location must be out of range of modern roots and other contaminants and should if possible be in a freshly cut exposure. Clean implements should be used and plastic gloves worn. When samples are required from cores a large-diameter slice of minimum vertical thickness should be taken. All samples should be dried individually at 85 °C and stored by double wrapping in polythene bags labelled on the outside with an indelible felt-tipped pen. No preservatives should be employed and a label should also be inserted between the two bags.

Table 12.3 Recommended optimal sample sizes for radiocarbon dating (after Worsley 1981). Note that if necessary it *may* be possible to date less material than these amounts.

Substance	Mass of raw sample (g)
wood	100–200
charcoal	50–100
peat	200–500
shells	75–100
bone	300–1000
soil organic matter	200–1000

Table 12.4 Recommended pre-treatments for water samples (based upon Brown et al. 1970).

Determination	Pre-treatment[†]	Determination	Pre-treatment[†]	Determination	Pre-treatment[†]
Al	2	N (nitrate)	1	acidity	3
B	1	N (nitrite)	1	alkalinity	3
Ba	2	N (organic)	4	CO_2 content	3
Ca	2	Na	1 or 2	colour	3
Cl	1	Ni	2	COD	4
Cu	2	P	1	hardness	1
F	1	Pb	2	pH	3
K	1 or 2	SiO_2	1	suspended solids	4
Mg	2	SO_4	1	total dissolved solids	1
Mn	2	Zn	2	total volatile solids	4
N (ammonia)	4			turbidity	4

[†]*Pre-treatments*: 1, field filtered, untreated; 2, field filtered, acidified; 3, unfiltered, settled; 4, unfiltered, well mixed.

12.4 Water sampling for chemical and physical analysis

Water samples should preferably be taken and transported in borosilicate glass (e.g. 'Pyrex') containers with close-fitting stoppers. The widely used polyethylene bottles are preferable to those of polypropylene and are essential if samples are frozen. Bottles should be filled completely unless they are to be frozen and should be agitated as little as possible. Bottles should be thoroughly cleaned before use and should be rinsed in the sampled waters immediately before sample collection. They should be labelled with details of date, site, time, water colour and temperature, discharge and discharge tendency, as appropriate. Field determination is mandatory for temperature, pH, dissolved oxygen content and specific conductance and is recommended for acidity and alkalinity. Other determinations may be made in the laboratory and should follow sample collection as quickly as possible. Recommended pre-treatments for samples after Brown *et al.* (1970) are given in Table 12.4. Filtration employs a 45 µm filter. Acidification is achieved by admixture of double-distilled reagent-grade HNO_3 to a resulting pH < 3. Analysis of samples with pre-treatment no. 3 should be made as soon as the container is opened and samples with pre-treatment no. 4 should be kept chilled if possible to deter bacterial activity. Other possible pre-treatments to reduce bacterial activity and to keep carbonates and iron in solution include freezing or acidification to pH < 1 with HCl or H_2SO_4. The latter may, however, bring colloids and fine particulate matter into solution. All samples should be kept out of direct sunlight and preferably stored in darkness. For further details of pre-treatments see Brown *et al.* (1970), which also gives recommendations concerning elements not included in Table 12.4.

13 Miscellaneous aids

A variety of aids and information are collected together in this final chapter. Section 13.1 is concerned with safety and gives advice on the minimum equipment for fieldwork, distress signals and emergency first aid. Since photographs are an indispensable means of recording phenomena the possible loss or malfunction of the user's exposure meter is catered for by the inclusion in Section 13.2 of a photographic exposure guide. Units and conversions between them are a constant headache and thus Section 13.3 deals briefly with the elements of the SI system and gives conversion factors between units of the SI, c.g.s. and Imperial systems. Many surveying and measurement exercises involve the use of simple trigonometry, and Section 13.4 gives a range of the more commonly used trigonometric formulae. Finally (Sec. 13.5), many tasks involve the estimation by eye of percentage cover or composition and Figure 13.2 may be used as an aid in such estimations.

13.1 Safety

Minimum recommended equipment for fieldwork away from the road or vehicle, and safety procedure

On safety grounds it is advised that the following equipment should be carried by all geomorphologists working away from vehicles, roads or habitation.

Waterproofs, with a hood, or other clothing appropriate to the climatic conditions likely to be experienced
first aid kit, with plasters, bandages, antiseptic cream and pain-killers
torch and whistle for signalling
emergency rations, e.g. glucose sweets
map and compass
spare warm clothing
survival bag or tent for shelter

Whenever possible work should not be carried out alone. Routes should be planned in advance and a copy of the routes should be left behind with someone who can take action should return be overdue. A weather forecast should be consulted before work if at all possible.

Equipment recommendations for expeditions is beyond the scope of this book, but much guidance can be found in Land (1978).

Distress signals

The international distress signal is the SOS signal in Morse code, i.e. three short, three long and three short calls, whistle blasts or flashes, in quick succession, and repeated at one minute intervals. An alternative is The Mountain Distress Signal in those countries where this is recognised; this consists of six long calls, whistle blasts or flashes in quick succession and repeated at one minute intervals. The Mountain Distress reply 'message understood' is three long calls, whistle blasts or flashes in quick succession, repeated at one minute intervals. It is not always advisable to make this signal, since it may cause those in distress to stop signalling before their exact position has been fixed.

Emergency first aid

The aims of emergency first aid are as follows:

1 Restore breathing⎫ these are the most important
2 Stop bleeding⎭
3 Bind up wounds and splint fractures
4 Make the injured person comfortable, provide shelter and care, and provide transport to medical facilities if possible.

These brief notes are no substitute for training in first aid, and provide only the most skeletal guidelines to action in the case of a very restricted range of commonly occurring injuries. For more detailed information see Turner (1974, 1975).

Exposure

Symptoms. Unexpected and seemingly unreasonable behaviour, complaints of cold and/or tiredness, physical and mental lethargy, slurring of the speech, excessive shivering, failure and/or abnormality in vision in extreme cases, violent outbursts of unexpected energy, falling or fainting fits, muscle cramp, ashen pallor.

Treatment. The patient is seriously ill. Provide immediate rest and shelter, insulation from further heat loss, especially under the body (vegetation, e.g. heather, is useful), hot sugar or glucose drink. If the patient stops breathing give him mouth-to-mouth respiration.

NEVER give alcohol, rub the limbs or apply *local* heat (e.g. hot water bottles). Do not allow the patient to walk. Send for help as soon as possible.

Exhaustion from heat, heatstroke

Symptoms. The patient becomes disorientated, often with severe headaches, feels extremely hot and often dry and has a rapid, thready pulse; may breath noisily and may lose consciousness. Also possibly lassitude, muscle cramp or vomiting.

Treatment. The patient is seriously ill; speed is essential. He must be cooled down as fast as possible. If you have plenty of water, pour it over him; if you can get near a stream, put him in it. If you have limited fluid only, wrap him in newspaper or other porous material and pour the liquid you have over him. Get the patient as soon as possible to where he can be put in a cold bath until his temperature comes down to normal levels.

Cold injury and frostbite

Symptoms. Early signs – tingling, stinging or aching in limb. Uncomfortable coldness followed by numbness. Skin red at first, waxy white later. Later signs – ranging from redness and swelling to the formation of blisters, or to complete destruction of the structure of the limb.

Treatment. Remove constricting clothing, frozen clothing being immersed in warm water if necessary. Increase circulation by applying warmth, e.g. by applying the hand to frozen parts. Maintain general body warmth and encourage sleep. If necessary put hot water bottles on the trunk. Cover large blisters with a loose dressing.

NEVER give alcohol, do not open blisters or apply ointments or lesions, and DO NOT overheat areas by hot water bottles or exposure to open fires. Do not apply massage or cold water soaks, and do not rub with snow or encourage the patient to walk around.

Fractures

Symptoms. If serious, are usually obvious. In other cases symptoms are shock (see below), tenderness, deformity and unnatural mobility, loss of movement, swelling and a sensation of grating between broken bone ends.

Treatment. Splint by improvised splints or by securing to other limbs or body to prevent movement. Treat for shock.

DO NOT attempt to set the fracture or to straighten the limbs, except when a long bone in the arm or leg is fractured. In these cases apply gentle traction to the limb by gently pulling it downwards before splinting. Do not attempt to force broken ends of bones back into wounds, and do not move the patient until the fracture has been splinted, unless his life is in danger.

Shock

Symptoms. Some shock is present in all cases of injury, and varies from a slight feeling of nausea to complete collapse. Symptoms are varied, but may include shallow, irregular and rapid breathing, dazed feeling, vomiting or feeling of sickness, cold clammy skin, dry mouth, pulse rapid and feeble, extreme pallor, enlarged pupils of eyes and even unconsciousness.

Treatment. Keep the patient warm, with AT ALL TIMES his feet slightly higher than his head. If necessary to transport unconscious patient down hill protect head adequately. Give warm sweet drinks if there is no stomach injury visible. Relieve pain if possible, loosen tight clothing, examine and treat slight injuries. Reassure the patient regularly.

Snake bite

Steps to take are as follows:

(1) Kill the snake and handle by the tail only. Keep for later identification.
(2) Reassure the victim. Death from snake bite is extremely rare in most countries and rare in all.
(3) Apply a ligature above the bite around the limb, to constrict the veins but not so tight as to constrict the arteries. Release this for 1 min in every 30 min – record these times.
(4) Douse the skin bitten with purified water. DO NOT cut the skin or suck the bite, use antiseptics or rub the skin.

(5) Treat the victim as a stretcher case if possible, wherever he has been bitten. Immobilise the limb as if it were fractured, and if possible cool it and keep it lower than the rest of the body.
(6) If there is pain give aspirin, codeine, etc., but NOT morphia. DO NOT administer Antivenin unless a highly skilled medical officer is present.
(7) Obtain medical aid as soon as possible, remembering to take the dead snake.

Scorpion bite

Scorpions are not designed to kill humans for their living. Except for very small children and the few who are hypersensitive to the venom, scorpions never kill people. The sting can be very unpleasant, but pain killers, and in particular local lignocaine anaesthetic, remove the pain and everything settles down when the limb is kept at rest. Prevention is better than cure; if you look where you tread before you get out of your bed in the morning and shake your shoes and trousers out, you will have no problems.

13.2   Photographic exposure guide

This guide (after Craven 1975) gives only an approximate indication of appropriate exposures and is intended for use only when more accurate methods of exposure setting are unavailable because of accident or oversight.

Set the shutter time equal to 1/ASA rating of the film (e.g. for ASA 125 film set the shutter speed at 1/125th of a second). Approximate conversion to ASA from DIN film speed number is given by Table 13.1. Use the following apertures for the conditions described:

Conditions	Aperture
bright or somewhat hazy sun; strong shadows	$f/16$
hazy sunshine; weak, poorly defined shadows	$f/11$
overcast but bright; no shadows visible	$f/8$
heavy overcast; thick fog, heavy rain, dark sky	$f/4$
open shade on sunny day (subject entirely in shade under open sky)	$f/5.6$

The settings above are applicable from about an hour after sunrise until about an hour before sunset. They must be modified for subjects darker or

Table 13.1 Approximate conversion to ASA from DIN film speed numbers (Feninger 1978).

ASA	DIN	ASA	DIN	ASA	DIN
800	30	200	24	50	18
640	29	160	23	40	17
500	28	125	22	32	16
400	27	100	21	25	15
320	26	80	20	20	14
250	25	64	19		

lighter than average, for backlighted subjects and for extreme high latitudes. Some guidance is given by the following:

for average subjects in light sand or snow use one f/stop smaller
for light coloured subjects use one f/stop smaller
for darker-than-average subjects use one f/stop larger
for backlit subjects use two f/stops larger
for high latitudes use one f/stop larger
for aerial photographs use one f/stop smaller.

Once a combination of shutter time and f/stop is chosen, any equivalent combination may also be used if it is wished to vary the exposure time, e.g. if 1/125th of a second and f/16 are suggested by the procedure above, 1/250th and f/11, or 1/60th and f/22 are also appropriate. Increasing the exposure time increases the depth of focus, decreasing the exposure time decreases the depth of focus.

13.3 Units and conversions

In the SI system of units, standard prefixes are used to denote multiples and submultiples of units and these are shown below:

Prefix	Symbol	Multiplier
tera-	T	10^{12}
giga-	G	10^{9}
mega-	M	10^{6}
kilo-	K	10^{3}
milli-	m	10^{-3}
micro-	μ	10^{-6}
nano-	n	10^{-9}
pico-	p	10^{-12}

The base units of the SI system can be combined to give derived units. The more commonly used derived units are given names and these together with their dimensions and equivalent c.g.s. units are given in Table 13.2. Note that the ratio in the final column is that of the actual unit. Thus the SI unit of velocity, the metre per second, is 100 times larger than the c.g.s. unit, the centimetre per second.

Table 13.3 gives conversion factors between commonly used SI, c.g.s. and Imperial units.

Table 13.2 Derived units in the SI and c.g.s. systems (Tennent 1971).

Quantity and recommended symbol	Dimensions	SI unit	c.g.s. unit	Ratio c.g.s.:SI units
mass, m	M	kilogram (kg)	gram (g)	10^{-3}
length, l	L	metre (m)	centimetre (cm)	10^{-2}
time, t	T	second (s)	second (s)	1
area, A, S	L^2	m^2	cm^2	10^{-4}
volume, V	L^3	m^3	cm^3	10^{-6}
density, ρ	ML^{-3}	$kg\ m^{-3}$	$g\ cm^{-3}$	10^3
velocity, u, v	LT^{-1}	$m\ s^{-1}$	$cm\ s^{-1}$	10^{-2}
acceleration, G	LT^{-2}	$m\ s^{-2}$	gal	10^{-2}
momentum, p	MLT^{-1}	$kg\ m\ s^{-1}$	$g\ cm\ s^{-1}$	10^{-3}
moment of inertia, I, J	ML^2	$kg\ m^2$	$g\ cm^2$	10^{-7}
angular momentum, L	ML^2T^{-1}	$kg\ m^2\ s^{-1}$	$g\ cm^2\ s^{-1}$	10^{-7}
force, F	MLT^{-2}	newton (N)	dyne (dyn)	10^{-5}
energy of work, E, W	ML^2T^{-2}	joule (J)	erg	10^{-7}
power, P	ML^2T^{-3}	watt (W)	$erg\ s^{-1}$	10^{-7}
pressure or stress, p	$ML^{-1}T^{-2}$	pascal (Pa)	$dyn\ cm^{-2}$	10^{-1}
surface tension, γ	MT^{-2}	$N\ m^{-1}$	$dyn\ cm^{-1}$	10^{-3}
viscosity, η	$ML^{-1}T^{-1}$	$kg\ m^{-1}\ s^{-1}$	poise	10^{-1}
frequency, ν, f	T^{-1}	hertz (Hz)	s^{-1}	1

Table 13.3 Conversion factors.

To convert A to B multiply by	A	B	To convert B to A multiply by
Length			
2.54×10^{-2}	inch (in)	metre (m)	39.37
0.3048	foot (ft)	metre (m)	3.2468
0.9144	yard (yd)	metre (m)	1.0936
1.8288	fathom (fm)	metre (m)	0.5468
20.1168	chain (ch)	metre (m)	4.97×10^{-2}
201.168	furlong (fl)	metre (m)	4.97×10^{-3}
1609.34	mile (mi)	metre (m)	6.214×10^{-4}
1853.2	nautical mile (UK)	metre (m)	5.3961×10^{-4}

Table 13.3 – *continued.*

To convert A to B multiply by	A	B	To convert B to A multiply by
1852.0	nautical mile (Int.)	metre (m)	5.3996×10^{-4}
1853.25	nautical mile (US)	metre (m)	5.3959×10^{-4}
100.0	cable	fathom (fm)	10^{-2}
1.6094	mile (mi)	kilometre (km)	0.6214
8.0	mile (mi)	furlong (fl)	0.125
1760.0	mile (mi)	yards (yd)	5.6818×10^{-4}
5280.0	mile (mi)	foot (ft)	1.8939×10^{-4}

Area

6.4516×10^{-4}	sq. inch (in^2)	sq. metre (m^2)	1.55×10^3
9.2903×10^{-2}	sq. foot (ft^2)	sq. metre (m^2)	10.764
0.8361	sq. yard (yd^2)	sq. metre (m^2)	1.196
2 589 988.0	sq. mile (mi^2)	sq. metre (m^2)	3.861×10^{-7}
4046.856	acre	sq. metre (m^2)	2.4711×10^{-4}
2.590	sq. mile (mi^2)	sq. kilometre (km^2)	0.3861
0.4047	acre	hectare (ha)	2.471
640.0	sq. mile (mi^2)	acre	1.5625×10^{-3}
10 000	hectare (ha)	sq. metre (m^2)	10^{-4}
3.861×10^{-3}	hectare (ha)	sq. mile (mi^2)	259.0045
247.105	sq. kilometre (km^2)	acre	4.047×10^{-3}

Volume

1.6387×10^{-5}	cubic inch (in^3)	cubic metre (m^3)	6.1024×10^4
16.387	cubic inch (in^3)	cubic centimetre (cm^3)	6.1024×10^{-2}
2.8317×10^{-2}	cubic foot (ft^3)	cubic metre (m^3)	35.314
28316.8	cubic foot (ft^3)	cubic centimetre (cm^3)	3.5315×10^{-5}
28.3168	cubic foot (ft^3)	litre (l)	3.5315×10^{-2}
0.7646	cubic yard (yd^3)	cubic metre (m^3)	1.3079
3.785×10^{-3}	gallon (US)	cubic metre (m^3)	264.2
4.546×10^{-3}	gallon (UK)	cubic metre (m^3)	219.97
0.21998	litre (l)	gallon (UK)	4.546
0.26418	litre (l)	gallon (US)	3.7853
8.0	bushel	gallon (UK)	0.125
9.608	bushel	gallon (US)	0.104
231.0	gallon (UK)	cubic inch (in^3)	4.329×10^{-3}
1233.482	acre foot	cubic metre (m^3)	8.1071×10^{-4}
0.8326	gallon (US)	gallon (UK)	1.2011
2.8413×10^{-2}	fluid ounce (fl. oz)	litre (l)	35.195

Mass

2.835×10^{-2}	ounce (oz)	kilogram (kg)	35.273
0.4536	pound (lb)	kilogram (kg)	2.2046
6.3503	stone (st)	kilogram (kg)	0.1575
50.8023	hundredweight (cwt)	kilogram (kg)	1.9684×10^{-2}
1016.04	ton	kilogram (kg)	9.8421×10^{-4}
907.20	short ton	kilogram (kg)	1.1023×10^{-3}
1.016	ton	metric tonne (t)	0.9842

Table 13.3 – *continued.*

To convert A to B multiply by	A	B	To convert B to A multiply by
2204.6	metric tonne (t)	pounds (lb)	4.536×10^{-4}
14.5939	slug	kilogram (kg)	6.852×10^{-2}
Weight			
4.448	pound force (lbf)	newton (N)	0.2248
0.1383	poundal (pdl)	newton (N)	7.233
10^5	newton (N)	dyne (dyn)	10^{-5}
32.17	pound-force (lbf)	poundal (pdl)	0.031 08
980.7	gram-force (gf)	dyne (dyn)	1.0197×10^{-3}
Pressure or stress			
15.44×10^6	ton-force per sq. in	pascal (Pa)	6.4767×10^{-8}
157.47	ton-force per sq. in	kilogram-force per sq. cm	6.3504×10^{-3}
107.3×10^3	ton-force per sq. ft	pascal (Pa)	9.3197×10^{-6}
1.0936×10^4	ton-force per sq. ft	kilogram-force per sq. m	9.1441×10^{-5}
6.895×10^3	pound-force per sq. in	pascal (Pa)	1.4503×10^{-4}
7.03×10^{-2}	pound-force per sq. in	kilogram-force per sq. cm	14.225
47.9	pound-force per sq. ft	pascal (Pa)	2.0877×10^{-2}
4.882	pound-force per sq. ft	kilogram-force per sq. m	0.204 83
101.325×10^3	standard atmosphere	pascal (Pa)	9.869×10^{-6}
1.033	standard atmosphere	kilogram-force per sq. m	9.6805
14.697	standard atmosphere	pound-force per sq. in	6.804×10^{-2}
760	standard atmosphere	millimetres of mercury (mm Hg)	1.315×10^{-3}
33.901	standard atmosphere	feet of water (ft H_2O)	2.9498×10^{-2}
3.05×10^{-2}	foot of water (ft H_2O)	kilogram-force per sq. cm	3.2787×10^{-3}
2.989×10^3	foot of water	pascal (Pa)	3.3456×10^{-4}
10^5	bar	pascal (Pa)	10^{-5}
133.322	millimetre of mercury	pascal (Pa)	7.5×10^{-3}
$9.806\ 65 \times 10^4$	kilogram-force per sq. cm	pascal (Pa)	1.0197×10^{-5}
Energy or work			
1.3558	foot pound-force	joule (J)	0.7376
1.3558×10^7	foot pound-force	erg	7.3757×10^{-8}
0.1383	foot pound-force	metre kilogram force	7.2307
4.2140×10^2	foot poundal	joule (J)	2.373×10^{-3}
4.2140×10^9	foot poundal	erg	2.373×10^{-10}
1.055×10^3	BTU	joule (J)	9.4787×10^{-4}
10^7	joule	erg	10^{-7}
4.1855	calorie at 15 °C	joule (J)	0.2389
Power			
550	horse-power	foot pound-force per second	1.8182×10^{-3}
7.457×10^9	horse-power	erg per second	1.341×10^{-10}
7.457×10^2	horse-power	watt (W)	1.341×10^{-3}
1.3405	kilowatt (kW)	horse-power	0.74599

Table 13.3 – *continued.*

To convert A to B multiply by	A	B	To convert B to A multiply by
1.3558	foot pound-force per second	watt (W)	0.7376
Density			
16.019	pound per cubic foot	kilogram per cubic metre	6.243×10^{-2}
1.6019×10^{-2}	pound per cubic foot	gram per cubic centimetre	6.243×10^{-3}
1.0012	ounces per cubic foot	gram per litre	0.9988
Unit weight			
16.019	pound-force per cubic foot	kilogram-force per cubic metre	6.243×10^{-2}
1.571×10^2	pound-force per cubic foot	newton per cubic metre	6.3654×10^{-3}
27.68	pound-force per cubic inch	gram-force per cubic centimetre	3.613×10^{-2}
271.4×10^3	pound-force per cubic inch	newton per cubic metre	3.6846×10^{-6}
Compressibility			
1.45×10^{-4}	sq. in per pound-force	sq. m per newton	6.897×10^3
14.22	sq. in per pound-force	sq. cm per kilogram-force	7.032×10^{-2}
9.324×10^{-6}	sq. foot per ton-force	sq. m per newton	1.0725×10^5
0.914	sq. foot per ton-force	sq. cm per kilogram-force	1.0941
Speed			
2.54	inch per second	centimetre per second	0.3937
30.48	foot per second	centimetre per second	3.281×10^2
0.447	mile per hour	metre per second	2.2371
0.5144	knot (Int.)	metre per second	1.944
1.0973	foot per second	kilometre per hour	0.9113
0.618 18	foot per second	mile per hour	1.61765
0.5925	foot per second	knot (Int.)	1.6878
3.6	metre per second	kilometre per hour	0.2778
0.9659×10^{-8}	foot per year	metre per second	1.0353×10^8
Rate of flow and discharge			
2832	cubic foot per second	cubic centimetre per second	3.53×10^{-4}
2.832×10^{-2}	cubic foot per second	cubic metre per second	35.311
76464	cubic yard per second	cubic centimetre per second	1.3078×10^{-5}
0.7646	cubic yard per second	cubic metre per second	1.3078
101.941	cubic foot per second	cubic metre per hour	9.8096×10^{-3}
2446.57	cubic foot per second	cubic metre per day	4.0874×10^{-4}
28.3161	cubic foot per second	litres per second	3.5316×10^{-2}
11.573 75	cubic metre per day	litres per second	8.6402×10^{-2}
0.408 735	cubic metre per day	cubic foot per second	2.4466
4.3813×10^{-2}	million gallons (US) per day	cubic metre per second	22.824

Table 13.3 – *continued.*

To convert A to B multiply by	A	B	To convert B to A multiply by
5.261×10^{-2}	million gallons (UK) per day	cubic metre per second	19.008
0.2713	acre-feet per day	million gallons (UK) per day	3.686
0.3259	acre-feet per day	million gallons (US) per day	3.0684
Yield			
0.699 725	cubic feet per acre	cubic metres per hectare	1.4291
1.120 85	pounds per acre	kilograms per hectare	0.892 18
Coefficient of consolidation			
0.1075	sq. in per minute	sq. cm per second	9.3023
1.075×10^{-5}	sq. in per minute	sq. m per second	9.3023×10^2
2.94×10^{-5}	sq. ft per year	sq. cm per second	3.4014×10^4
2.94×10^{-9}	sq. ft per year	sq. m per second	3.4014×10^8
Concentration			
1.0012×10^{-3}	ounce per cubic foot	gram per cubic centimetre	9.988×10^2
1.0012	ounce per cubic foot	kilogram per cubic metre	0.9988
16.019×10^{-3}	pound per cubic foot	gram per cubic centimetre	62.426
16.019	pound per cubic foot	kilogram per cubic metre	6.2426
Dynamic viscosity			
47.8803	pound-seconds per square foot	newton-seconds per sq. metre	2.0885×10^{-2}
47.8803	slugs per foot-second	newton-seconds per sq. metre	2.0885×10^{-2}
10^{-3}	centipoise (cP)	newton-seconds per sq. metre	10^3
Kinematic viscosity			
$9.290\ 30 \times 10^4$	sq. foot per second	centistoke (cSt)	1.0764×10^{-5}
$9.290\ 30 \times 10^{-2}$	sq. foot per second	sq. metre per second	10.764
10^{-6}	centistoke (cSt)	sq. metre per second	10^6

13.4　Simple trigonometric relations

Notation is shown in Figure 13.1 and applies to both right-angled and oblique triangles.

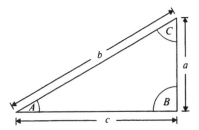

Figure 13.1 Notation for right-angled and oblique triangles. For right-angled triangles $B = 90°$.

Right angled triangles

$\sin A = a/b = \cos C$
$\cos A = c/b = \sin C$
$\tan A = a/c = \cot C$
$\cot A = c/a = \tan C$
$a = b \sin A = b \cos C = c \tan A = c \cot C$
$b = a/\sin A = a/\cos C = c/\cos A = c/\sin C$
$c = b \cos A = b \sin C = a \tan A = a \cot C$
$C = 90° - A$
$c = (a^2 + b^2)^{1/2}$
$a = \{(b + c)(b - a)\}^{1/2}$

Oblique triangles

$a/\sin A = b/\sin B = c/\sin C$
$a = b \sin A/\sin B = c \sin A/\sin C$
$c = a \sin C/\sin A$
$b = a \sin(A + C)/\sin A$
$b = a \sin B/\sin A$
$a = c \sin A/\sin C$
$s = \tfrac{1}{2}(a + b + c)$
$c = (a^2 + b^2 - 2ab \cos C)^{1/2}$
$\sin \tfrac{1}{2}A = \{(s - b)(s - c)/bc\}^{1/2}$
$\cos \tfrac{1}{2}A = \{s(s - a)/bc\}^{1/2}$
$\tan \tfrac{1}{2}A = \{(s - b)(s - c)/s(s - a)\}^{1/2}$
area $= \{s(s - a)(s - b)(s - c)\}^{1/2} = \tfrac{1}{2}ab \sin C$
$\qquad\qquad\qquad = a^2 \sin B \sin C/2 \sin A.$

13.5 Estimation of percentage cover

An aid to the estimation of percentage cover or composition is shown in
Figure 13.2.

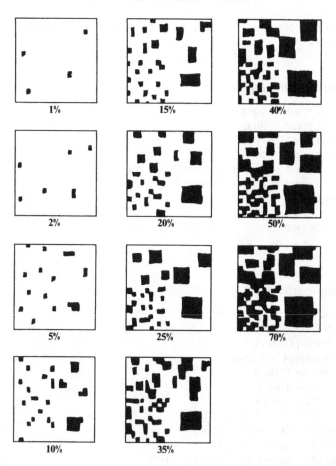

Figure 13.2 Chart for estimating percentage cover or composition (after Hodgson
1974).

References

Ackers, P., W. R. White, J. A. Perkins and A. J. M. Harrison 1978. *Weirs and flumes for flow measurement.* Chichester: Wiley.

Allen, J. R. L. 1963. The classification of cross-stratified units with notes on their origin. *Sedimentology* 2, 93–114.

Allen, J. R. L. 1965. A review of the origin and characteristics of recent alluvial sediments. *Sedimentology* 5, 91–191.

Allen, J. R. L. 1968. *Current ripples.* Amsterdam: North Holland.

Allen, J. R. L. 1969. On the geometry of current ripples in relation to stability of fluid flow. *Geogr. Annlr* 51A, 61–96.

Armstrong, T., B. Roberts and C. Swithinbank 1966. *Illustrated glossary of snow and ice.* Cambridge: The Scott Polar Research Institute/Unesco.

Bagnold, R. A. 1941. *The physics of blown sand and desert dunes.* London: Methuen.

Bagnold, R. A. 1954. *The physics of blown sand and desert dunes*, 2nd edn. London: Methuen.

Bagnold, R. A. 1956. The flow of cohesionless grains in fluid. *Phil. Trans R. Soc. A* 249, 235–97.

Bagnold, R. A. 1963. Beach and nearshore processes. In *The sea*, Vol. III, M. N. Hill (ed.), 664–9. New York: Wiley.

Bagnold, R. A. 1980. An empirical correlation of bedload transport rates in flumes and natural rivers. *Proc. R. Soc. A* 372, 453–73.

Baker, V. R. and D. F. Ritter 1975. Competence of rivers to transport coarse bedload material. *Geol. Soc. Am. Bull.* 86, 975–8.

Barnes, H. H. 1967. *Roughness characteristics of natural channels.* US Geological Survey Water Supply Paper no. 1849.

Benson, M. A. 1968. *Measurement of peak discharge by indirect methods.* Technical Note 90. Publication of WMO 225, TP1/9. Geneva: World Meteorological Organisation.

Berg, D. W. 1968. Systematic collection of beach data. In *Proceedings of the 11th Conference on Coastal Engineering, September 1968, London*.

Bhattacharya, P. K. and H. P. Patra, 1968. *Direct current geoelectrical sounding.* Amsterdam: Elsevier.

Bieniawski, Z. T. 1975. The point load test in geotechnical practice. *Q. J. Engng Geol.* 9, 1–11.

Bishop, A. W. and N. R. Morgenstern 1960. Stability coefficients for earth slopes. *Géotechnique* 15, 1–34.

Blachut, T. J. and F. Müller 1966. Some fundamental considerations on glacier mapping. *Can. J. Earth Sci.* 3, 747–9.

Black, J. N. C., C. W. Bonython and J. A. Prescott 1954. Solar radiation and the duration of sunshine. *Q. J. R. Met. Soc.* **80**, 231–5.

Blong, R. J. 1972. Methods of slope profile measurement in the field. *Aust. Geog. Stud.* **10**, 182–92.

Boersma, R. 1975. Rock types and sedimentary structures. In *Tidal deposits* R. N. Ginsburg (ed.), endpapers. Berlin: Springer Verlag.

Bogardi, J. L. 1961. Some aspects of the application of the theory of sediment transportation to engineering problems. *J. Geophys. Res.* **66**, 3337–46.

Bowles, J. E. 1979. *Physical and geotechnical properties of soils*. New York: McGraw-Hill.

Bray, D. J. 1979. Estimating average velocity in gravel-bed rivers. *Proc. Am. Soc. Civ. Engrs J. Hydraul. Div.* **105**, **HY-9**, 1103–22.

Bretschneider, C. L. 1966. In *Shore protection, planning and design*, 3rd edn. Technical Report no. 4, 50–114. Coastal Engineering Research Center.

Brewer, R. 1964. *Fabric and mineral analysis of soils*. New York: Wiley.

British Geomorphological Research Group (various dates). *BGRG Technical Bulletins* 1–25. Norwich: Geo Abstracts.

British Museum (Natural History) 1970. *Fossils, minerals and rocks*. Instructions for collectors no. 11. London: British Museum (Natural History).

Broch, E. and J. A. Franklin 1972. The point-load strength test. *Int. J. Rock Mech. Min. Sci.* **9**, 669–97.

Brown, E., M. W. Skougstad and M. J. Fishman 1970. *Methods for collection and analysis of water samples for dissolved minerals and gases*. Techniques of Water Resources Investigations of the United States Geological Survey, Book 5. Washington, DC: US Government Printing Office.

Bruno, R. O. and L. W. Hiipakka 1973. Littoral Environment Observation program in the state of Michigan. In *Proceedings, 16th Conference on Great Lakes Research*, 492–507. International Association of Great Lakes Research.

Brunsden, D. 1971. Ever-moving hillsides. *Geog. Mag.* **43**, 759–64.

BS 812 1975. *Methods of sampling and testing of mineral aggregates sands and fillers. Part 2. Physical properties*. London: British Standards Institution.

BS 1377 1967. *Methods of testing soils for civil engineering purposes*. London: British Standards Institution.

BS 1377 1975. *Methods of test for soils for civil engineering purposes*. London: British Standards Institution.

BS 3680 1964. *Methods of measurement of liquid flow in open channels. Part 3A. Velocity area methods listed*. London: British Standards Institution.

BS 3680 1965. *Methods of measurement of liquid flow in open channels. Part 4A. Weirs and flumes – thin plate weirs and venturi flumes*. London: British Standards Institution.

BS 3680 1973. *Methods of measurement of liquid flow in open channels. Part 8A. Current meters incorporating a rotating element*. London: British Standards Institution.

BS 3680 1980. *Methods of measurement of liquid flow in open channels. Part 3A. Velocity area methods*, 2nd edn. London: British Standards Institution.

BS 3681 1973. *Methods for testing sampling and testing of lightweight aggregates for concrete*. London: British Standards Institution.

Cailleux, A. 1947. L'indices dè émoussé des grains de sable et grès. *Rev. Geomorph. Dynam.* **3**, 78–87.

Campbell, C. V. 1967. Lamina, laminaset, bed and bedset. *Sedimentology* **8**, 7–26.

Carpenter, E. W. 1955. Some notes concerning the Wenner configuration. *Geophys. Prospect.* **3**, 388–402.

Carson, M. A. 1971. Application of the concept of threshold slopes to the Laramie Mountains, Wyoming. In *Slopes*, D. Brunsden (ed.), Trans. Inst. Br. Geogs, Spec. Publ. no. 3, 31–48.

Carson, M. A. and M. J. Kirkby 1972. *Hillslope form and process*. Cambridge: Cambridge University Press.

Chepil, W. S. and N. P. Woodruff 1957. Sedimentary characteristics of dust storms II. Visibility and dust concentration. *Am. J. Sci.* **255**, 104–14.

Cherven, V. B. 1978. Fluvial and deltaic facies in the Sentinal Butte Formation, Central Williston Basin. *J. Sed. Petrol.* **48**, 159–70.

Chow, V. T. 1959. *Open-channel hydraulics*. New York: McGraw-Hill.

Chowdhury, R. N. 1978. *Slope analysis*. Amsterdam: Elsevier.

Church, M. 1975. *Electrochemical and fluorometric tracer techniques for streamflow measurements*. BGRG Technical Bulletin 12. Norwich: Geo Abstracts.

Churchill, R. R. 1979. A field technique for profiling precipitous slopes. In *Shorter technical methods (III)*. BGRG Technical Bulletin 24, 29–34. Norwich: Geo Abstracts.

Classen, D. F. 1977. Temperature profiles for the Barnes Ice Cap surge zone. *J. Glaciol.* **18**, 391–406.

Clifton, H. E. 1976. Wave-formed sedimentary structures: a conceptual model. In *Beach and nearshore sedimentation*, R. A. Davis & R. L. Ethington .(eds), 126–48. Tulsa: Society of Economic Paleontologists and Mineralogists.

Colby, B. R. and D. W. Hubbell 1961. *Simplified method for computing total sediment discharge with the modified Einstein method*. US Geological Survey Water Supply Paper no. 1593.

Collins, J. I. 1976. Approaches to wave modeling. In *Beach and nearshore sedimentation*, R. A. Davis & R. L. Ethington (eds), 54–68. Tulsa: Society of Economic Paleontologists and Mineralogists.

Compagnie Général de Géophysique 1963. *Abaques de sondage électrique, Master curves for electrical sounding*. European Association of Exploration Geophysicists.

Compton, R. R. 1962. *Manual of field geology*. New York: Wiley.

Cooke, R. U. and J. C. Doornkamp 1974. *Geomorphology in environmental management*. Oxford: Oxford University Press.

Cooke, R. U. and A. Warren 1973. *Geomorphology in deserts*. London: Batsford.

Coulomb, C. A. 1776. *Essais sur une application des règles des maximis et minimis à quelques problems de statique relatifs à l'architecture*. Mémoirs présentées par divers Savants. Academie des Sciences Paris.

Cousins, B. F. 1978. Stability charts for simple earth slopes. *Proc. Am. Soc. Civ. Engrs, J. Geotech. Engng Div.* **104**, **GT-4**, 267–79.

Cowan, W. L. 1956. Estimating hydraulic roughness coefficients. *Agric. Engng, Lond.* **37**, 473–5.

Craven, G. M. 1975. *Object and image.* Englewood Cliffs, NJ: Prentice-Hall.

Culmann, C. 1866. *Graphische Statik.* Zürich.

Curtis, L. F. and S. Trudgill 1975. *The measurement of soil moisture.* BGRG Technical Bulletin 13. Norwich: Geo Abstracts.

Dapples, E. C. and J. F. Rominger 1945. Orientation analysis of fine grained clastic sediments: a report of progress. *J. Geol.* **53**, 246–61.

Darbyshire, M. and L. Draper 1963. Forecasting wind-generated sea waves. *Engineering* **195**, 482–4.

Day, M. J. and A. S. Goudie 1977. Field assessment of rock hardness using the Schmidt test hammer. In *Shorter technical methods (II)*, BGRG Technical Bulletin 18, 19–29. Norwich: Geo Abstracts.

Deere, D. U. and R. P. Miller 1966. *Engineering classification and index properties for intact rock.* Air Force Weapons Laboratory Technical Report AFWL-TR-65-116. Kirkland Base, New Mexico.

Demek, J. (ed.) 1972. *Manual of detailed geomorphological mapping.* Prague: Academia.

Demek, J. and C. Embleton (eds) 1978. *Guide to medium-scale geomorphological mapping.* Brno: International Geographical Union.

de Quervain, M. R. 1966. On avalanche classification, a further contribution. In *Proceedings of the international symposium on scientific aspects of snow and ice avalanches (Davos, 1965).* Int. Assoc. scient. Hydrol. Publ. no. 69, 410–7.

Dobrin, M. B. 1976. *Introduction to geophysical prospecting.* New York: McGraw-Hill.

Dury, G. H. 1976. Discharge prediction, present and former, from channel dimensions. *J. Hydrol.* **30**, 219–45.

Edwards, A. M. C., A. T. McDonald and J. R. Petch 1975. *The use of electrode instrumentation for water analysis.* BGRG Technical Bulletin 15. Norwich: Geo Abstracts.

Einstein, H. A. 1942. Formulas for the transportation of bed-load. *Trans Am. Soc. Civ. Engrs* **107**, 561–74.

Einstein, H. A. 1950. *The bedload function for sediment transportation in open channel flows.* Technical Bulletin no. 1026. US Department of Agriculture.

Ethridge, F. G. and S. A. Schumm 1978. Reconstructing paleochannel morphologic and flow characteristics: methodology, limitations and assessment. In *Fluvial sedimentology*, A. D. Miall (ed.), 703–22. Calgary: Canadian Society of Petroleum Geologists.

Ewing, M., G. P. Woollard and A. C. Vine 1939. Geophysical investigations in the emerged and submerged Atlantic Coastal Plain, Part 3, Barnegat Bay, New Jersey section. *Geol Soc. Am. Bull.* **50**, 257–96.

Faegri, K. and H. Gams 1937. Entwicklung and Vereinheitliehung der Signaturen fur Sediment-und-Torfarten. *Geol. Stockh. Forhdl.* **59**, 273.

Feininger, A. 1978. *The complete photographer*. London: Thames and Hudson.

Finlayson, B. 1979. Electrical conductivity: a useful technique in teaching geomorphology. *J. Geog. higher Educ.* **3**, 68–87.

Fleming, N. C. 1965. Form and function of sedimentary particles. *J. Sed. Petrol.* **35**, 381–90.

Folk, R. L. 1954. The distinction between grain size and mineral composition in sedimentary rock nomenclature. *J. Geol.* **62**, 344–59.

Folk, R. L. 1968. *Petrology of sedimentary rocks*, 2nd edn. Texas: Hemphills.

Folk, R. L. 1974. *Petrology of sedimentary rocks*, 3rd edn. Texas: Hemphills.

Freyberger, S. G. and G. Dean 1979. Dune forms and wind regime. In *A study of global sand seas*, E. D. McKee (ed.), 137–70. Washington, DC: US Geological Survey.

Galvin, C. J. 1967. Longshore current velocity: a review of theory and data. *Rev. Geophys.* **5**, 287–304.

Galvin, C. J. 1968. Breaker-type classification on three laboratory beaches. *J. Geophys. Res.* **73**, 3651–9.

Garde, R. J. and K. G. Ranga Raju 1977. *Mechanics of sediment transportation and alluvial stream problems*. New Delhi: Wiley Eastern.

Gardiner, V. and R. V. Dackombe 1977. A simple method for the field survey of slope profiles. In *Shorter technical methods (II)*, BGRG Technical Bulletin 18, 9–18. Norwich: Geo Abstracts.

Geological Society Engineering Group Working Party 1972. The preparation of maps and plans in terms of engineering geology. *Q. J. Engng Geol.* **5**, 295–382.

Geological Society Engineering Group Working Party 1977. The description of rock masses for engineering purposes. *Q. J. Engng Geol.* **10**, 355–88.

Glennie, K. W. 1970. *Desert sedimentary environments*. London: Elsevier.

Goudie, A. (ed.) 1981. *Geomorphological techniques*. London: George Allen and Unwin.

Graf, W. H. 1971. *Hydraulics of sediment transport*. New York: McGraw-Hill.

Greeley, R. C., J. D. Iversen, J. B. Pollack, N. Udovich and B. White 1974. Wind tunnel studies of Martian aeolian processes. *Proc. R. Soc. A* **341**, 331–60.

Gregory, K. J. 1979. Changes of drainage network composition. *Acta Univ. Ouluensis A* **82**, 19–28.

Griffiths, D. H. and R. F. King 1965. *Applied geophysics for engineers and geologists*. Oxford: Pergamon.

Gubler, H. U. 1975. *On the Rammsonde hardness equation*. IAHS-AISH Publ. no. 114, 110–21.

Hagedoorn, J. G. 1959. The plus–minus method of interpreting seismic refraction sections. *Geophys. Prospect.* **7**, 158–83.

Harms, J. C. and R. K. Fahnestock 1965. Stratification, bed forms and flow phenomena (with an example from the Rio Grande). In *Primary sedimentary structures*, G. V. Middleton (ed.), 84–115. Tulsa: Society of Economic Paleontologists and Mineralogists.

Harrison, W. D. 1975. Temperature measurements in a temperate glacier. *J. Glaciol.* **14**, 23–30.

Harvey, J. G. 1976. *Atmosphere and ocean: our fluid environments*. Sussex: Artemis Press.

Hawkins, L. V. and D. Maggs 1961. Nomograms for determining maximum errors and limiting conditions in seismic refraction surveys with a blind zone problem. *Geophys. Prospect.* **9**, 526–32.

Henderson, F. M. 1966. *Open channel flow*. New York: Macmillan.

Hey, R. D. 1979. Flow resistance in gravel-bed rivers. *Proc. Am. Soc. Civ. Engrs, J. Hydraul. Div.* **105**, **HY-4**, 365–79.

Hjulstrom, F. 1935. Studies of the morphological activity of rivers as illustrated by the River Fyris. *Bull. Geol Inst. Univ. Uppsala* **25**, 221–57.

Hodgson, J. M. (ed.) 1974. *Soil survey field handbook*. Soil Survey Technical Monograph no. 5. Harpenden, Herts: Rothamsted Experiment Station.

Hoek, E. and J. W. Bray 1977. *Rock slope engineering*, 2nd edn. London: Institute of Mining and Metallurgy.

Holdsworth, G. 1975. Measurement of small strain-rates over short time periods. *J. Glaciol.* **14**, 317–24.

Holmes, P. 1975. Wave conditions in coastal areas. In *Nearshore sediment dynamics and sedimentation*, J. Hails & A. Carr (eds), 1–16. London: Wiley.

Hooke, R. Le B. and B. R. Koci 1978. Temperature measurements on the Barnes Ice Cap, Baffin Island, Canada. *J. Glaciol.* **20**, 441–2.

Hunter, R. E. 1977. Basic types of stratification in small eolian dunes. *Sedimentology* **24**, 361–87.

Hutchinson, J. N. 1968. Mass movement. In *Encyclopaedia of geomorphology*, R. W. Fairbridge (ed.), 688–96. New York: Halsted.

Hvorslev, M. J. 1960. Physical components of the shear strength of saturated clay. In *Proceedings of the ASCE research conference on shear strength of cohesive soils*, 1–8. New York: American Society for Civil Engineering.

Ingram, R. L. 1954. Terminology for thickness of stratification and parting units in sedimentary rocks. *Geol Soc. Am. Bull.* **65**, 937–8.

Inman, D. L. 1963. Ocean waves and associated currents. In *Submarine geology*, F. P. Shepard (ed.), 48–81. New York: Harper and Row.

Inman, D. L. and R. A. Bagnold 1963. Littoral processes. In *The sea*, vol. III, M. N. Hill (ed.), 529–53. New York: Wiley.

Jakosky, J. J. 1957. *Exploration geophysics*. California: Trija.

Jansen, P. L. 1979. *Principles of river engineering*. London: Pitman.

Jones, R. L. and P. R. Cundill 1978. *Introduction to pollen analysis*. BGRG Technical Bulletin 22. Norwich: Geo Abstracts.

Jopling, A. V. 1966. Some principles and techniques used in reconstructing the hydraulic parameters of a paleo-flow regime. *J. Sed. Petrol.* **36**, 5–49.

Jopling, A. V. and R. G. Walker 1967. Morphology and origin of ripple-drift cross lamination with examples from the Pleistocene of Massachussetts. *J. Sed. Petrol.* **38**, 971–84.

Keller, G. V. and F. C. Frischknecht 1966. *Electrical methods in geophysical prospecting*. Oxford: Pergamon.

Kesel, R. H. 1976. The use of refraction seismic techniques in geomorphology. *Catena* **3**, 91–8.

King, C. A. M. 1972. *Beaches and coasts*, 2nd edn. London: Edward Arnold.

Kirkpatrick, W. W. 1965. Effects of grain size and grading on the shearing behavior of granular materials. *Proceedings of the 6th International Conference on soil mechanics and foundation engineering* **1**, 273–7. Toronto: Univ. of Toronto Press.

Kottlowski, F. E. 1965. *Measuring stratigraphic sections*. New York: Holt, Rinehart and Winston.

Krumbein, W. C. 1941. Measurement and geological significance of shape and roundness of sedimentary particles. *J. Sed. Petrol.* **11**, 64–72.

Lahee, F. H. 1931. *Field geology*, 3rd edn. New York: McGraw-Hill.

Lahee, F. H. 1961. *Field geology*, 6th edn. New York: McGraw-Hill.

Land, A. E. 1978. *The expedition handbook*. London: Butterworth.

Limerinos, J. T. 1970. *Determination of the Manning coefficient from measured bed roughness in natural channels*. Water Supply Paper no. 1898-B. US Geological Survey.

Lohnes, R. A. and R. L. Handy 1968. Slope angles in friable loess. *J. Geol.* **76**, 247–58.

Longuet-Higgins, M. S. 1952. On the statistical distribution of the heights of sea waves. *J. Mar. Res.* **11**, 245–6.

McKee, E. D. 1966. Structure of dunes at White Sands National Monument, New Mexico (and a comparison with structures of dunes from other selected areas). *Sedimentology* **7**, 3–69.

McKee, E. D. (ed.) 1979. *A study of global sand seas*. US Geological Survey Professional Paper no 1052.

McKee, E. D. and G. W. Weir 1953. Terminology of stratification and cross stratification. *Geol Soc. Am. Bull.* **64**, 381–90.

Mamayev, O. I. 1975. *Temperature–salinity analysis of world ocean waters*. Amsterdam: Elsevier.

Meidav, T. 1960. Nomograms to speed up seismic refraction computations. *Geophysics* **25**, 1035–53.

Meyer-Peter, E. and R. Muller 1948. Formulas for bed-load transport. In *Proceedings of the 3rd meeting of the International Association for Hydraulics Research, Stockholm*. 39–64.

Mooney, H. M. and W. W. Wetzel 1956. *The potentials about a point electrode and apparent resistivity curves for a two-, three- and four-layer earth*. Minneapolis: University of Minnesota Press.

Neville, A. M. 1973. *Properties of concrete*. London: Pitman.

Nijman, W. and C. Puigdefabregas 1978. Coarse-grained point bar structure in a molasse-type fluvial system, Eocene Castisent Formation, South Pyrenean Basin. In *Fluvial sedimentology*, A. D. Miall (ed.), 487–510. Calgary: Canadian Society of Petroleum Geologists.

Nye, J. F. 1959. A method of determining the strain-rate tensor at the surface of a glacier. *J. Glaciol.* **3**, 409–19.

O'Brien, M. P. 1942. *A summary of the theory of oscillatory waves*. Technical Report no. 2, 1–14. Beach Erosion Board, US Army Corps of Engineers.

Open University 1978. *Oceanography – Sediments*, S334 Course Book. Milton Keynes: Open University.

Orellana, E. and H. M. Mooney 1966. *Master tables and curves for vertical electrical sounding over layered structures*. Madrid: Interciencia.

Østrem, G. and A. D. Stanley 1966. *Glacier mass balance measurements. A manual for field work*. Ottawa: Department of Mines and Technical Surveys.

Ouma, J. P. B. M. 1967. Fluviatile morphogenesis of roundness. The Hacking River, New South Wales, Australia. In *River Morphology*. Int. Assoc. Sci. Hydrol. Publ. no. **75**, 319–44.

Parasnis, D. S. 1979. *Principles of applied geophysics*. London: Cahpman and Hall.

Paterson, W. S. B. 1969. *The physics of glaciers*. Oxford: Pergamon.

Patrick, D. A. and R. L. Weigel 1955. Amphibian tractors in the surf. In *First conference on ships and waves*, 397–422. London and New York: The English Foundation Council on Wave Research and the American Society of Naval Architects and Marine Engineers.

Pitty, A. F. 1966. Some problems in the location and delimitation of slope profiles. *Z. Geomorph.* **10**, 454–61.

Pitty, A. F. 1968. A simple device for the field measurement of hillslopes. *J. Geol.* **76**, 717–20.

Powers, M. C. 1953. A new roundness scale for sedimentary particles. *J. Sed. Petrol.* **23**, 117–9.

Redpath, B. B. 1973. *Seismic refraction exploration for engineering site investigations*. Report A.D. 768 710. US Department of Commerce, National Technical Information Service.

Reineck, H. E. and F. Wunderlich 1968. Classification and origin of flaser and lenticular bedding. *Sedimentology* **11**, 99–104.

Rijkwaterstaat, The Netherlands 1980. *Standard graphs for resistivity prospecting*. European Association of Exploration Geophysicists.

Riley, N. A. 1941. Projection sphericity. *J. Sed. Petrol.* **11**, 94–7.

Rivière, A. and P. Ville 1967. Sur l'utilisation d'une indice morphologique nouveau dans la representation d'une formation detritique grossiere. *C.R. Hebd. Séanc. Acad. Sci. Paris D* **265**, 1369–72.

Roberts, A. 1977. *Geotechnology*. Oxford: Pergamon.

Robinson, E. 1968. Effects of air pollution on visibility. In *Air pollution*, vol. I, A. C. Stern (ed.), 349–400. New York: Academic Press.

Robinson, G. W. 1949. *Soils, their origin, constitution and classification*, 3rd edn. London: Murby.

Rogers, J. K. 1947. Temperature corrections in altimeter surveying. *Bull. Am. Assoc. Petrolm Geol.* **31**, 371–4.

Russel, R. C. H. and D. H. Macmillan 1952. *Waves and tides*. London: Hutchinson.

Rutley, F. and H. H. Read 1970. *Rutley's elements of mineralogy*, 26th ed. London: George Allen & Unwin.

Savigear, R. A. G. 1965. A technique of morphological mapping. *Ann. Assoc. Am. Geogs* **53**, 514–38.

Schneiderhohn, P. 1954. Eine vergleichende studies uber methoden zur quantitativen bestimmung von abrundung und form an sandkornern. *Heidelberger Beitr. Min. Petrol.* **4**, 172–91.

Scott, R. F. 1963. *Principles of soil mechanics*. Reading, Mass: Addison-Wesley.

Selby, M. J. 1980. A rock mass strength classification for geomorphic purposes: with tests from Antarctica and New Zealand. *Z. Geomorph.* **24**, 31–51.

Seligman, G. 1936. *Snow structure and ski fields*. London: Macmillan.

Sen, N. 1978. A chart for optimisation of sample size. *Indian J. Earth Sci.* **5**, 214–6.

Sharpe, C. F. S. 1938. *Landslides and related phenomena*. New York: Columbia University Press.

Shaw, C. F. 1928. A definition of terms used in soil literature. In *Proceedings and Papers of the 1st International Congress on Soil Science*, vol. 5, 38–64.

Shepard, F. P. 1954. Nomenclature based on sand–silt–clay ratios. *J. Sed. Petrol.* **24**, 151–8.

Shields, I. A. 1936. Anwendung de Ahnlichkeitmechanik und des turbulenz for schung auf die geschiebebewegung. *Mitt. Preuss. VersAnst. Wasserbau Erd Schiffbau* **26**. English translation by W. P. Ott and J. C. van Uchelen, California Institute of Technology, Passadena.

Simons, D. B. and E. V. Richardson 1962a. Resistance to flow in alluvial channels. *Trans Am. Soc. Civ. Engrs* **127**, 927–1006.

Simons, D. B. and E. V. Richardson 1962b. *The effect of bed roughness on depth discharge relations in alluvial channels*. Water Supply Paper no. 1498E. US Geological Survey.

Simons, D. B. and E. V. Richardson 1971. Flow in alluvial sand channels. In *River mechanics*, Vol. I, H. W. Shen (ed.), 9.1–9.89. Fort Collins: privately published by H. W. Shen.

Simons, D. B. and F. Şentürk 1976. *Sediment transport technology*. Fort Collins: Water Resources Publications.

Singer, D. H. and S. Yaalon 1974. Vertical variation in strength and porosity of calcrete (nari) on chalk, Shefela, Israel and interpretation of its origin. *J. Sed. Petrol.* **44**, 1016–23.

Skempton, S. W. 1945. Earth pressure and the stability of slopes. In *The principles and application of soil mechanics*. London: Institution of Civil Engineers.

Skempton, A. W. 1964. The long-term stability of clay slopes. *Géotechnique* **14**, 75–102.

Sneed, E. D. and R. L. Folk 1958. Pebbles in the lower Colorado River, Texas: a study in particle morphogenesis. *J. Geol.* **66**, 114–50.

Southard, J. B. 1971. Representation of bed configurations in depth–velocity–size diagrams. *J. Sed. Petrol.* **41**, 903–15.

Sternberg, R. W. 1972. Predicting initial motion and bedload transport of sediment particles in the shallow marine environment. In *Shelf sediment transport*, D. J. P. Swift, D. B. Duane & O. H. Pilkey (eds), 61–82. Stroudsburg: Dowden, Hutchinson and Ross.

Stone, R. O. and H. J. Summers 1972. *Study of subaqueous and subaerial sand ripples*. Final Report no. USC-Geology-72-1. Department of Geological Sciences, University of Southern California.

Strickler, A. 1923. *Beitrage zur Frage der Geschwindigheits-formel und der Rauhigkeitszahlen für Strome, Kanale und Geschlossene Leitungen*. Mitteilungen des Eidgenössischer Amtes fur Wasserwirtschaft no. 16, Bern.

Sundborg, A. 1967. Some aspects on fluvial sediments and fluvial morphology. I. General views and graphic methods. *Geog. Ann.* **49A**, 333–43.

Tagg, G. F. 1934. Interpretation of resistivity measurements. *Trans Am. Inst. Min. Metal. Engrs* **110**, 135–47.

Tanner, W. F. 1967. Ripple mark indices and their uses. *Sedimentology* **9**, 89–104.

Taylor, G. and K. D. Woodyer 1978. Bank deposition in suspended-load streams. In *Fluvial sedimentology*, A. D. Miall (ed.), 257–76. Canadian Society of Petroleum Geologists Memorandum no. 5. Calgary: Canadian Society of Petroleum Geologists.

Telford, W. M., L. P. Geldart, R. E. Sheriff and D. A. Keys 1976. *Applied geophysics*. Cambridge: Cambridge University Press.

Tennent, R. M. (ed.) 1971. *Science data book*. Edinburgh: Oliver and Boyd.

Terzaghi, K. 1943. *Theoretical soil mechanics*. New York: Wiley.

Terzaghi, K. and R. B. Peck 1967. *Soil mechanics in engineering practice*. New York: Wiley.

Thomas, R. H. 1976. The distribution of 10 m temperatures on the Ross Ice Shelf. *J. Glaciol.* **16**, 111–7.

Turner, A. C. 1974. *The traveller's health guide*. Newton Abbot: David and Charles.

Turner, A. C. 1975. *Travel medicine – a handbook for practitioners*. London: Longman.

Unesco/IASH/WMO 1970. *Seasonal snow cover*. Technical Paper in Hydrology No. 2.

US Department of Agriculture 1951. *Soil survey manual*. Agricultural Handbook no. 18. US Department of Agriculture.

Van Nostrand, R. G. and K. L. Cook 1966. *Interpretation of resistivity data*. US Geological Survey Professional Paper no. 499.

Varnes, D. J. 1958. Landslide types and processes. In *Landslides and engineering practice*, E. B. Eckel (ed.). Special Report no. 29, Highway Research Board. 20–47.

Wagg, C. J. and K. Echelmeyer 1979. Rhombus and rhomboid parallelogram patterns on glaciers: natural indications of strain. *J. Glaciol.* **22**, 247–61.

Wadell, H. 1933. Sphericity and roundness of rock particles. *J. Geol.* **41**, 310–31.

Wadell, H. 1935. Volume, shape and roundness of quartz particles. *J. Geol.* **43**, 250–80.

Waters, R. S. 1958. Morphological mapping. *Geography* **43**, 10–19.

Wentworth, C. K. 1919. A laboratory and field study of cobble abrasion. *J. Geol.* **27**, 507–21.

Wentworth, C. K. 1922. A scale of grade and class terms for clastic sediments. *J. Geol.* **30**, 377–92.

Wentworth, C. K. 1933. The shapes of rock particles: a discussion. *J. Geol.* **41**, 306–9.

Williams, G. 1964. Some aspects of the eolian saltation load. *Sedimentology* **3**, 257–87.

Wilson, I. G. 1970. *The external morphology of wind-laid sand deposits*. PhD thesis. University of Reading.

Wilson, I. G. 1972. Aeolian bedforms – their development and origins. *Sedimentology* **19**, 173–210.

Worsley, P. 1981. Radiocarbon dating: principles, application and sample collection. In *Geomorphological techniques*, A. Goudie (ed.), 277–83. London: George Allen and Unwin.

Young, A. 1972. *Slopes*. Edinburgh: Oliver and Boyd.

Young, A., with D. Brunsden and J. B. Thornes 1974. *Slope profile survey*. BGRG Technical Bulletin 11. Norwich: Geo Abstracts.

Zeller, J. 1963. Einfuhrung in den Sedimenttransport offener Gerinne. *Schweiz. Bauzeitung* 81D.

Zingg, Th. 1935. Beiträge zur Schotteranalyse. *Min. Petrog. Mitt. Schweiz.* **15**, 39–140.

Index

Italicised entries refer to text figures, and bold entries to text sections.